The Story of Nic Volker

and the Dawn of

Genomic Medicine

One in
a Billion

Mark Johnson
and Kathleen Gallagher

SIMON & SCHUSTER

New York   London   Toronto   Sydney   New Delhi

To our families,
Mary-Liz Shaw and Evan Johnson and
Bob, Andrew, and Emily Nicol: For their sustaining love
and faith through the years it took to complete this book

Simon & Schuster
1230 Avenue of the Americas
New York, NY 10020

First Simon & Schuster hardcover edition April 2016

SIMON & SCHUSTER and colophon are registered trademarks
of Simon & Schuster, Inc.

For information about special discounts for bulk purchases,
please contact Simon & Schuster Special Sales at 1-866-506-1949
or business@simonandschuster.com.

The Simon & Schuster Speakers Bureau can bring authors to your
live event. For more information or to book an event contact the
Simon & Schuster Speakers Bureau at 1-866-248-3049 or visit
our website at www.simonspeakers.com.

Interior design by Lewelin Polanco

Manufactured in the United States of America

1   3   5   7   9   10   8   6   4   2

Library of Congress Cataloging-in-Publication Data is available

ISBN 978-1-4516-6132-3
ISBN 978-1-4516-6134-7 (ebook)

Contents

One in
a Billion

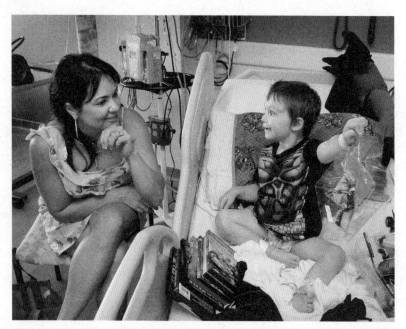

Nic Volker and his mother, Amylynne

Chapter 1

At the Threshold

I t may turn out that the roots of love—the mystical force that makes a mother fight to save her sickly son, that drives a doctor to the limits of medicine to keep the boy alive, that stirs scientists to take the first, uncertain steps into a new age in the search for a cure—it may be that this deep, instinctual caring can be traced to one of the smallest, most essential parts of us: the long, pale, winding thread of chemicals deep inside our cells. Our genetic script. Our personal code. Our DNA.

It was a century and a half ago that the Austrian monk Gregor Mendel famously experimented with pea plants, showing how they pass physical traits from one generation to the next. His research provided the first insight into why some of us are short or tall, brown-eyed or blue-eyed, hairy or bald. In the many decades since, scientists have pinned these traits to genes, distinct chunks of DNA that together tell the story of all we carry with us into our brief life on earth: the risks we face, the quirks we display, the diseases we may be destined to endure.

To what extent we're the sum of our genes, we cannot say. Scientists are still in the early stages of exploring the genome. But with each passing year, they move ever closer to the fundamental mysteries behind the genes we inherit and how they interact with our environment and upbringing to shape us as human beings.

James Watson and Francis Crick, who discovered DNA's double helix, did not foresee such grand inquiries into the origins of our traits and diseases when they began their quest to find the secret of life. For them, "the secret," as Crick called it, was simply the structure of DNA, the chemical form in which all of Mendel's traits were spelled out and stored within our cells. When they discovered the double helix in 1953, it was the answer to a chemistry puzzle. They never considered that it might one day be possible to read the code.

"We didn't think ahead to that," Watson recalled one afternoon in the fall of 2011, sitting behind his desk at Cold Spring Harbor Laboratory. "Neither of us speculated that it would be nice if we could sequence DNA. We weren't chemists. We never had any feeling as to when it would occur."

Yet their work established a foundation upon which others built. In 1977, British and American scientists devised the first techniques for reading the genetic code. And at the turn of this century, a massive effort involving hundreds of researchers using hundreds of complex machines succeeded in unveiling the entire sequence of our DNA, the human genome. The ability to read our genome has led humans to a place that Mendel, Watson, and Crick could scarcely have imagined, a place that forces us to confront profound questions of fate and chance. Gains in computing speed and capacity are bringing the cost down so rapidly that it is inevitable that our descendants will have their DNA sequenced at birth.

After decades of debate, the ethical issues and fears surrounding these changes are crossing from hypothetical to real. And at the *forefront is a new form of medicine—one that seeks to make sense of human complexity, to sift through thousands of harmless differences in the genetic script, to find the errors and defects that threaten our survival, and to use this new knowledge to treat patients in need.*

Pioneers of this nascent medicine—a remarkable team of doctors and scientists and a family with a very sick child—came together in Wisconsin in the summer of 2009. Their story has reverberated around the globe, and it won't be long before most of us find that, to one degree or another, our lives are shaped by what our doctors find in that genetic script.

So where does the script begin?

Imagine a tiny camera, smaller and more powerful than any now in existence, small enough to pass through skin, deep into the human body, into the blood. The camera slips inside one of the body's basic units, a single white blood cell. Then it penetrates the cell's dark nucleus. There lie the twenty-three matched pairs of chromosomes that contain our essence, the DNA that makes us who we are and determines our risks for cancer, heart disease, and many other illnesses that can change the course of our lives.

Risks force us to confront our limited time on earth, to make the decisions that come to define us. Should we finally visit the Sistine Chapel or the Great Pyramid at Giza? Is it time to change careers? Should we renounce hamburgers, exercise more, and take cholesterol-lowering drugs, all for a better chance at a longer life?

Until very recently the risks inscribed in our genes were inaccessible to us. Today, with companies such as 23andMe allowing us to glimpse a snapshot of the script, we are entering an age in which our genes may come to possess an almost Delphic power over us.

Our imaginary camera moves onward, deeper into the blood cell, into the nucleus, into a chromosome. The chromosome consists of hundreds of genes, each one a basic unit of heredity, a set of instructions for brown eyes, red hair, or male-pattern baldness. The camera passes into one of these genes and travels along the two coiled strands of DNA. The strands form the double helix, a structure resembling two side-by-side, winding tape measures that pass through all of the chromosomes. Linking the two strands are 3.2 billion rungs, like steps on a ladder.

At each rung sit two chemical bases. There are only four possibilities for the bases: cytosine, guanine, adenine, or thymine, commonly reduced to the shorthand C, G, A, or T.

The camera stops, focusing on one of the two letters on a single rung, the tiniest portion of one gene. A point among 3.2 billion points inside the nucleus of a blood cell. A place so small that it must be magnified seven million times using a powerful electron microscope in order to be seen.

It could be anyone's blood cell. But it isn't.

This lonely milepost rests inside the body of a four-year-old boy who loves Batman and squirt-gun fights and bouncing on his hospital bed. He loves steak, but he seldom gets to eat it. Most of the time, the boy's food comes in liquid form, seeping into his bloodstream through an intravenous line. Doctors have been forced to take desperate measures to save the boy. Holes, tiny pencil pricks, have been forming inside the child's intestines whenever he eats real food, a life-threatening condition.

No one knows the reason for the havoc inside the child's intestine. It is a mystery doctors have been struggling to solve for two years and counting, but to no avail. Without an answer they can barely keep up with his symptoms. How can they fight a disease they cannot explain?

They have no camera to scan the double helix.

They do not know what secrets it holds.

⁓

What a terrible disease, thinks Amylynne Santiago Volker, as she waits with her son in the holding area outside the operating room. Other families sit nearby, but it is obvious from their nervous eyes that being here is different for them—an unusual experience, frightening and new. For Amylynne and four-year-old Nic there is nothing new about the trip to the operating room at Children's Hospital of Wisconsin. By early 2009, the trips have become routine. Routine to a degree most could not imagine.

Today's trek to the operating room is just one in a series of more than a hundred for Nic, a never-ending exercise in futility that began more than two years earlier when his mother found a strange abscess on his bottom. Since then his intestine has been ravaged by tiny holes called fistulas, which form after he eats food; this is why so much of his nutrition must come through an intravenous line. The holes cause stool to drain onto his abdomen, raising the constant threat of infection. Nic's disease has doctors so baffled that they cannot come up with a way to stop new holes from sprouting. The best they can do is try to stave off infections. So almost every day, Amylynne sits with Nic until it's time for the anesthesiologist to put him to sleep; then the boy's surgeon, Marjorie Arca, cleans the same wound she has cleaned dozens of times before.

Nic, a small boy with bright blue eyes and a toddler's high-pitched voice, waits to begin his day in the operating room the way other children wait to begin theirs in day care or kindergarten. In the holding area on this June morning, Nic is the only kid who appears dressed for Halloween.

His blue eyes stare out from a Batman mask. His body, so small for his age, is cloaked in a Batman cape. His hands wear Batman gloves that make the sound *Ka-pow!* He marches around the hospital as if it were his backyard. In the midst of a marathon stay that will exceed 250 days, the hospital has, in essence, replaced home.

Amylynne, a striking, dark-haired woman of Filipino descent, is in her early forties and carries herself like none of the other parents in the hospital. Like them, she has cried this morning. But she has already cleaned up her face and reapplied her makeup to hide her worry from Nic. As many times as she walks her son to the operating room, her anxiety about the anesthesia that will put him under never wanes. Unlike other parents, she has learned to hide it. Unlike other parents, she knows the receptionists by name, the social workers, the nurses. She knows whether they have boyfriends or husbands, whether they have children, whether they jog or work out at the gym.

Nothing about Amylynne is passive, not anymore. She monitors every aspect of her son's care and decides who should be on his team, not just doctors but nurses, too. She searches the Internet and medical reports for clues to Nic's disease and possible treatments. Raised in a medical family and schooled by her son's illness, she has gained enough confidence in her hunches to share them with the doctors. When she disagrees with their decisions, she tells them. She is determined to be taken seriously; the doctors have no idea how far she will go. She will change how she looks, how she acts. She will argue with staff and make enemies if she has to. Her devotion to Nic, her love, makes her more formidable than any other parent these doctors have encountered.

Some of the staff know that this mother is virtually living at the hospital. Some have seen her curl up beside Nic at night. She's not supposed to sleep with her son, but she wants to be close and there are few opportunities in the hospital. The nurses are not about to tell this woman that she cannot sleep beside the son she may soon lose. What amazes the nurses is how, after sitting beside her son late into the night, Amylynne manages to get up, style her hair, and put on a fresh business suit as if headed to another day at the office. If the hospital has become Nic's de facto home, his health has become Amylynne's twenty-four-hour-a-day job. Everything else in her life seems trivial by comparison.

What the other parents in the holding area and the doctors and nurses do not see is the cumulative toll that so many trips to the operating room have taken. They do not see the entries in her journal, agonizing over each new disappointment, each failure to find a cause, each new hole the doctors find.

"Inflammation, inflammation, inflammation," she writes in her journal, marveling at the sublime cruelty of the disease. "If you nourish Nic with real food, even the formula—the food causes bacteria in the gut and then you have inflammation—which causes the fistulas and wound sites, etc., etc., etc. What a vicious cycle."

Nic craves food. Pizza. Steak. In the hospital, he falls asleep some nights clutching a bag of Bagel Bites the way other children cuddle teddy bears.

Amylynne must play the killjoy. She tells her son it may be a long time before he eats normal food, then listens to him howl in frustration, this little boy who is already severely underweight.

The disease punishes him for eating and punishes him when he doesn't.

When she is not cataloging the insults of his illness, Amylynne, a born-again Christian, fills her online journal with affirmations of her faith. "God is good all the time," she writes, though the words sometimes sound like more of a reminder to herself. Often she prays with Arca, the surgeon, or with Nic. When the stabbing pain in his abdomen grows unbearable, Nic asks her to pray for him.

An entire routine has evolved around Nic's near-daily trips to the operating room. He selects music. He picks a flavor of oxygen. He holds the mask through which the oxygen flows. If residents try to hold the mask, he pushes their hands away. Others control most of what goes on in his world, poking needles into his body, taking blood whenever they please, taking his temperature when all he wants is to play. What little control he has, even over something as trivial as holding an oxygen mask, he clings to.

The day's routine begins with a simple act that encapsulates for Amylynne the pain and frustration of these last two years.

When it is time, when the doctor is ready, Amylynne carries Nic to the operating room. She stops at a line marked on the floor, the point she cannot cross without wearing scrubs, the point where she must leave her son with the doctors. Here she is forced to entrust his fragile life to modern medicine.

But in Nic's case, medicine appears to have reached a similar line. Centuries of progress toward understanding heredity and genes have left doctors at a threshold. When a patient has a known genetic illness, there is often a test that can confirm it. And in some cases, scientists have figured out how to fix the defect or compensate for it.

However, when doctors have spoken to Amylynne, all they have been able to do is tell her the diseases Nic does not have. They have run tests on individual genes and on his immune system. They have said that his problem resembles known illnesses. Yet it hasn't responded to known treatments.

Lately the message from doctors has been that he appears to have a new disease, something no one has ever seen. Something found nowhere in the millions of entries in medical journals. If the illness is new, there is no test to confirm it. There is no understanding of the cause and no clear picture of what is happening inside Nic. Why does the simple act of eating—something we all do, something we *must* do to survive—cause Nic such terrible pain and put his life in peril? The doctors do not understand the problem they are trying to fix.

So Nic and his family have set out on what is known in medicine as a diagnostic odyssey. They have gone from one expert to another, one theory to another—and now they are back at Children's Hospital of Wisconsin, where they have spent the majority of Nic's illness, where the doctors know them and the staff refuses to give up.

Amylynne is tired, but she has not exhausted all hope; her capacity for hope seems endless. Someone, she knows, will figure out her son's disease in time to save him. Somewhere there is an answer.

It is a strong conviction, but not blind faith. Many of her relatives are doctors. She knows that medicine is ever advancing, exploring new ideas.

Among scientists, excitement has been building around one of those ideas. Ever since the completion of the Human Genome Project's first draft in 2001, researchers have been seeking to apply knowledge of the genome to human health. This is the next frontier: the chance to bring genomics into the realm of medicine, to treat real patients who have puzzling diseases and nowhere else to turn.

Thousands of human diseases are caused by a single mutation in an individual gene, and the Human Genome Project has raised, for the first time, the possibility of sequencing all 21,000 or so of a person's genes. If it works, gene sequencing could transform medicine and alter millions of lives. It would allow scientists to hunt for mutations on an unheard-of scale. So far researchers have traced the origins of some five thousand rare diseases to specific flaws in our genetic script. They're still hunting for the defects responsible for another two to three thousand single-gene disorders. And while they may prove more difficult to treat, complex, multigene disorders are also yielding some of their secrets to the technology of genetic sequencing. The ability to scan the genome now brings them all within the grasp of modern medicine.

Imagine a single tool that could pinpoint the causes of so much suffering. These rare mutations and diseases are only rare when viewed individually. Taken together, rare diseases, most of them hereditary, afflict 25 million to 30 million Americans, roughly one in ten.

Nic's hopes, like theirs, may hinge upon the promise of genomic medicine. This boy, though unknown to the hundreds of scientists working to harness the power of our genetic script, may present the very opportunity they have been seeking: to translate the mammoth genome project into medicine.

Without knowing it, the Volkers have come to American

medicine at a crucial time. Pioneering scientists have been anticipating, even staking their careers on, the arrival of a child like Nic,
someone with a new disease that requires a new medicine. There
is no way to know when the moment will come; doctors and scientists can only hope that when it does, they are ready.

Much is at stake. Medical breakthroughs require risk and can
release unintended consequences. For every new treatment that
succeeds, others fail, setting back the advance of science and,
worse, harming patients.

But for Nic and Amylynne, there is little comfort in the con-
ventional methods that keep failing against an unconventional
disease. Where scientists see risk, Amylynne sees possibility, the
chance to escape their grim routine.

At the line marking the operating room's entrance, Amylynne
waits to pass her boy to the surgeon and a medical system that is
so far mystified. She hugs and kisses Nic and tells him good-bye.

Then she hands over her only son.

Chapter 2

Beyond the Four Letters

APRIL 1993

More than fifteen years before Nic's disease surfaced, scientists had begun a quest to unlock the human genome, to turn it from a vaguely defined map of inheritance to a medical tool. The effort began in 1990, when the US government made a crucial commitment: hundreds of millions of dollars toward the goal of decoding the entire genetic script.

Among the scientists who entered the field were three unlikely participants, physiologists who ran into one another at a medical conference in Southern California in the spring of 1993. The genome-sequencing effort was still in its early stages, but the physiologists saw the promise of a new era in medicine and were eager to stake out a place in it. Their meeting on a beach in La Jolla set into motion an ambitious plan that would lead a small midwestern medical school to upend the traditional pecking order and take a pioneering role in bringing genomic medicine to patients.

During a break at the 10th Scientific Meeting of the Inter-American Society of Hypertension, the three scientists stepped outside. They had planned on a brief chat, but it quickly grew into a much deeper discussion. Conspicuous in their suits and ties, they now walked along the beach. The men paid little attention to the Pacific Ocean, glistening under a clear blue sky. They had bigger things in mind: they were discussing genes.

The talk ebbed and flowed, from quiet conversation to loud debate and back. Just three years into the massive Human Genome Project, they imagined a bold future. One day, computers would be powerful enough to analyze thousands of genes. It would then be up to scientists to match those genes to specific functions in the body. A comprehensive understanding of how genes affect the body's many systems could change the course of medicine. The knowledge would free doctors to delve deep into the workings of a terrible disease, halt its progression, maybe even prevent it from happening in the first place.

The men were not trained geneticists; instead they saw the body from an all-encompassing perspective. They were trained in physiology, the branch of biology that studies how organs, tissues, cells, and molecules work together to keep living systems alive and functioning. Still, they shared a vision for how to proceed with the new science of genomics.

Allen Cowley, the oldest by twenty years, was a thin man with large glasses. An established scientist with expertise in hypertension, he headed the physiology department at the Medical College of Wisconsin. Cowley had known the other two, Howard Jacob and José Krieger, both in their early thirties, for some time. Krieger, the son of a prominent Brazilian physiologist, had studied under Cowley at the Medical College. Jacob had done his graduate work at the University of Iowa under Cowley's friend Michael Brody, a pharmacology professor. But when Brody died unexpectedly of a heart attack while attending a conference in

Australia, Cowley stepped in to become Jacob's mentor, his scientific father.

Like Cowley, the younger scientists were experts in cardiovascular physiology, with a keen understanding of the system by which blood carries oxygen, nutrients, and other materials throughout the body. While researchers in other areas of science were embracing the trend toward breaking things down into the smallest units, the physiologists understood that there was no minimizing what happens in the body. Even the simple act of standing up involves a series of changes in blood flow, blood pressure, nerve response, and heart rate—all important.

This awareness, they believed, enhanced their view of the scientific shift under way.

The march toward decoding the body's genetic script had been steady and painstaking. In 1953, James Watson and Francis Crick, relatively obscure researchers at the time, discovered DNA's double-helix structure, an advance that enhanced the understanding of how human inheritance worked. In 1977, the Americans Walter Gilbert and Allan Maxam and the Englishman Frederick Sanger discovered two different ways to read the sequence, to identify which of the four chemical bases—adenine, guanine, cytosine, or thymine (A, G, C, or T)—lies at a specific point on each rung of the double helix. The entire set of instructions still seemed too vast to read, some 3.2 billion pairs of letters. Then, in the mid-1980s, Sanger's technique was automated and hope emerged that the entire human sequence could be mapped.

When the US government launched the Human Genome Project in 1990, announcing that for the next fifteen years it would provide an estimated $200 million a year for the effort, young scientists steered their careers toward the great question of the day: DNA.

The new field brimmed with possibility. If researchers could solve the mystery of how genes translate heredity into chemicals that course through the body, it might be possible to control the process. Here might lie the power to drastically improve the length and quality of human life.

The scientists on the beach came well prepared to meet the challenges of the new era. Two of them had worked for leaders in the fast-growing field.

After receiving their schooling in the Midwest, Jacob and Krieger had both moved to Boston for postdoctoral fellowships under Victor Dzau, a cardiologist who would later serve as the first editor of a journal conceived by Cowley called *Physiological Genomics*. Dzau would go on to become president of the National Academy of Sciences' Institute of Medicine. Under his tutelage, the young fellows had begun learning the language and techniques of genomics as the science took shape before their eyes.

While in Dzau's lab, Jacob also undertook a second fellowship, this one with Eric Lander, of the Whitehead Institute for Biomedical Research at MIT. Lander, a mathematician turned geneticist, had been one of the key figures in convincing the US Congress to commit billions of dollars to the genome project. He would be the lead author on the landmark 2001 paper in the journal *Nature* that presented a first draft of the sequence—an early benchmark against which other sequences would be measured.

It was clear in 1993—eight years before that draft was presented—that this project would be *the* crucial event in biology in the last half of the twentieth century, arguably as significant to humanity as putting a man on the moon.

Many researchers in physiology and other disciplines were defensive, critical of the vast amount of resources being devoted to the genome. They worried that all of the money going into one area would mean that there would be less for their own work. The project, they said, did not represent "real" science. It turned the centuries-old scientific method on its head; researchers would be searching a massive data set for answers rather than developing a hypothesis and testing it.

But Cowley and his young protégés didn't see it that way. They believed that this effort to draft the human blueprint would fuel research and discoveries for years to come.

None of it seemed difficult on that beautiful day in La Jolla. The afternoon slipped away, and the conference sessions they had left behind came to a close as the scientists continued their walk. Cowley emphasized over and over that molecular biology and physiology had to merge; that physiologists had a necessary part of the tool kit for mining genetic information. As the men talked, ideas took shape and plans formed. They could build upon work Jacob was doing with rats to find the genes responsible for disease in the animals; then they could translate that work to humans.

It was a new scientific world they were entering. Cowley felt reinvigorated. He was certain that no other physiologists were strategizing this way, preparing for what would come to be known as personalized medicine. Not even leaders in the field were as well positioned. Dzau understood what was happening, but he did not have direct connections to a physiology group that was working with animals. And he was busy in a new position as chief of the division of cardiovascular medicine at Stanford University. Lander was increasingly occupied with the Human Genome Project.

Meanwhile, the disciplines of physiology and genomics were moving on separate paths. Physiologists, accustomed to traditional measures such as blood pressure, pulse, and heart rate, were not learning the new ways of genomics. The sequencing experts were so intent on the complicated technical job of decoding the genome that they weren't thinking about tying sequence to function.

That left wide opportunities for Cowley and the two young scientists as they looked toward the future.

The Human Genome Project would use as many as eight hundred sequencing machines to determine the order of the chemical base pairs on the double helix. For this knowledge to be useful, though, the sequence would need to be linked to the vast catalogue of processes in the body: the way the kidneys regulate our balance of salt and water, the factors that make a blood vessel relax and constrict.

Cowley knew those functions as well as anyone. He was the son of a well-known cardiologist who was chief of medicine at the Harrisburg Polyclinic Hospital. From an early age, he followed in his father's footsteps, dissecting hearts in the basement of his family's Pennsylvania home. His father said they were cows' hearts. Cowley had his doubts.

During a two-year fellowship in the Jackson, Mississippi, lab of Arthur Guyton, Cowley had grown more convinced of the importance of physiology. The field provided a means for understanding the body's complex systems. Even as science was shifting its focus from broad systems to their tiny building blocks, Guyton's lab was building a new model of how the cardiovascular system works.

Cowley, Guyton's protégé, spent hours in the lab during the late 1960s and early 1970s twisting calibrated knobs, called potentiometers, that lined the drawers of an old analog computer. Guyton had purchased the computer at the local army surplus store to help fine-tune equations that modeled the workings of biological systems. Although he was not named on the scientific paper, Cowley's work contributed to the 1972 publication of the Guyton cardiovascular model, a pioneering mathematical description of the body's circulatory controls. The model looked like an electric circuit diagram, showing connections among blood flow, oxygen delivery, and other aspects of the circulatory system. It revolutionized how scientists thought about the way the heart and other organs work.

Cowley stayed with Guyton for twelve years and rose rapidly to the rank of professor, leaving only for a one-year sabbatical at Harvard University. There he found certain academic areas stifled by tradition. He was disappointed with the overall caliber of Harvard's physiology department. Yet Cowley made the most of his stay. He made friends outside the department, including Dzau, and built connections that would help his future recruiting efforts.

Years later, Cowley would still view Guyton as the most impressive scientist he ever met. But by 1980, Cowley's wife, Terrie, was anxious to leave Jackson. Nine years earlier, she had been deeply involved in the unsuccessful gubernatorial campaign of Charles Evers, the brother of the slain civil rights activist Medgar Evers. Now the civil rights movement was waning, and she longed for new opportunities. Cowley, too, was eager for more responsibility. He felt ready to run his own physiology department.

Despite offers from bigger, better-known institutions, he chose the Medical College of Wisconsin. After decades of financial difficulties, the Medical College had lost the sponsorship of Marquette University in 1967. Newly autonomous and riding a wave of successful fund-raising, the school had begun to grow. It had recently constructed a new building on the outskirts of Milwaukee near several existing hospitals.

Still, when Cowley arrived in 1980, the physiology department had just two faculty members engaged in high-level, federally funded work. The department had been without a chairman for two years. He saw it as an empty shell offering him the opportunity to build a unique research enterprise. Cowley figured he could unite the trend toward minimalism with his focus on the big picture, connect the microscopic world to the whole systems at work in the human body. He would then recruit young faculty with backgrounds in molecular biology, and later bioengineering, computer modeling, and genomics, to work with the physiologists on challenging problems.

Dean Edward Lennon was not offering a lot of money, but it was enough. And what was lacking in funds, Lennon made up for in encouragement.

"Academia places great stock upon name, and I had already gotten over this shallow view of excellence, having been in Jackson,

Mississippi, for more than a decade," Cowley said. "It was clear to me that if Dr. Guyton could become the world's most preeminent physiologist, working in the so-called armpit of the nation, that Milwaukee would not in itself be a handicap that I could not overcome."

By that afternoon on the beach in 1993, Cowley had spent more than a decade at the Medical College, putting his imprint on both the department and the school. He had gotten himself appointed to nearly every search committee, whether for faculty members, deans, or the Medical College president. While physiology departments at other schools withered, his grew.

As the science of genomics emerged, he saw how it could transform physiology. But he knew he would need help to launch that transformation.

A well-worn phrase, used by Isaac Newton and many others, holds that accomplished researchers stand on the shoulders of giants. Cowley knew he stood on the shoulders of Arthur Guyton and those who had gone before him. Now, to build on Guyton's work in complex systems, Cowley needed a protégé: the next scientist in the chain.

Howard Jacob didn't know it yet, but Cowley was homing in on him.

Chapter 3

Try Me

1993–1996

Allen Cowley would not have tried to hire Howard Jacob in the late 1980s, when the young researcher was writing his dissertation examining how the brain regulates blood pressure. Frankly, Cowley had become bored with such traditional topics. But over the last five years, Jacob had grown as a scientist. He was now working at the cutting edge of genomics, part of an elite group pioneering an emerging field. He knew the landscape and the techniques.

This young man, with his easy smile and tassel of brown hair glancing across his forehead, possessed an important quality: he took risks. He wanted to pursue trailblazing science.

Jacob had become interested in the field when his father, a navy engineer, was stationed on Midway Island, a speck of land in the middle of the vast Pacific. Five-year-old Howard loved to peer into the water, watching the fish nibble at the bait on his line. Early in life, he decided to become a marine biologist.

That goal would evolve over time, but science remained his passion. The boy read *National Geographic* from cover to cover and convinced his parents to break the rule of no television during dinner so that he could watch Jacques Cousteau. Jacob marveled at the way Cousteau could breathe underwater, see things never seen before, and share them with the world.

Darlene Strysik, Jacob's sixth-grade teacher, stoked his interest by finding a high school physics teacher to help him develop a science fair project on water wave behavior. The project was chosen for a statewide competition. Jacob went on to get a biology degree at Iowa State University in Ames, Iowa, and spent summers working with scientists at Abbott Laboratories, near his parents' home. One summer, Jacob administered drugs to the pharmaceutical company's rats and screened them for arthritis. The next, he worked in the cardiovascular lab of the drug developer Jaroslav Kyncl.

His experiences led him into the field of pharmacology, which became the focus of his PhD at the University of Iowa in Iowa City. By the time he finished there, he had a strong grounding in how human and rat bodies worked. Once a hated carrier of contagious diseases, the rat had become an indispensable tool in experimental medicine and drug development, increasingly used to find the causes of human disease. Rats are especially useful for studying the cardiovascular system, which is exactly what Jacob wanted to do.

It was Eric Lander, more than anyone else, who would help him combine that knowledge with genomics. When Jacob arrived at Lander's lab in 1989, researchers were using mice for most experiments, just as they were at the other lab Jacob worked in, that of Victor Dzau. But Jacob soon convinced Lander to focus more on rats. During Jacob's four years in the lab, Lander taught him how to work with genes—and, even more importantly, how to think big.

Lander, who ran the Whitehead Institute for Biomedical Research at MIT, viewed DNA with a numerical bias. The genome was a three-billion-letter text file, a little less than a gigabyte of information. Three-letter DNA segments, called codons, make up the words of the file; there are 4^3, or 64, possible codons. The codons serve as instructions for making twenty different amino acids. Chained together, amino acids combine to form proteins, the workhorses that carry out functions such as digesting food, clotting blood, and contracting muscles. The whole process—from assembling the smallest chunks of DNA to making the proteins that direct breathing and other actions—is so rife with numbers and patterns, it was made for a mathematician.

Lander knew that interpreting DNA's long script would require sifting through massive amounts of data. This would change biology into a computational science, relying heavily on algorithms and other formulas to read all of the new information.

Early on, Jacob explained to Lander that he wanted to find the genetic basis of high blood pressure in rats, a goal that would be a major step toward understanding high blood pressure in humans. It was a bold, pie-in-the-sky goal at a time when the rat genome had barely been explored. And Lander's response surprised him.

"Do you know how many chromosomes a rat has?" Lander asked.

"No."

"Wouldn't that be a good thing to know if you were interested in the genetics of high blood pressure?"

Lander prodded Jacob's ambition instead of curbing it, and Jacob took the challenge to heart. In Lander's lab, everyone was fighting to get somewhere. Jacob came to think of them all as a "merry band of misfits."

In fact, Jacob was more of a misfit than the others, a midwestern rat physiologist in an East Coast lab full of molecular biologists and geneticists. He may have lacked certain credentials, but he possessed a ferocious work ethic.

Jacob would spend three days a week in Dzau's lab and four in Lander's, then flip the schedule the following week. His days started about 9:00 a.m. and ended about 11:00 p.m. On the way home, he stopped to pick up his wife, Lisa, from her second-shift job as a medical technologist.

"There was a reason we were married nine years without children," she would later say.

One of Jacob's jobs in Lander's lab was to find what are called markers, generally meaningless variations of DNA that occur in reliable places on the genome. At a time when no one knew the order of DNA's letters, the markers were the only way to home in on the areas of the genome that controlled selected traits.

In the 1980s, Lander attracted attention by using markers as the basis for a new DNA mapping technique he developed with David Botstein, an MIT genetics professor who would become vice president of science at Genentech, one of the early giants of biotechnology.

The logic behind their detective work went back to Gregor Mendel, the nineteenth-century monk considered to be the father of modern genetics. Mendel spent about eight years in the mid-1800s cross-fertilizing some 28,000 pea plants to determine patterns that dictated the inheritance of certain characteristics. He discovered the basic principles of heredity, the rules explaining how traits such as stem length and pea-pod color are passed from one generation to the next. Scientists like Lander could use those rules to trace the inheritance of a given trait and of the markers that were passed along with it. Because the markers' positions on specific chromosomes were known, researchers could use them to figure out the general location of the genes that controlled specific traits.

Lander and five collaborators used the technique to conduct, in 1989, the first-ever survey of an entire genome—that of a tomato—which successfully found the regions controlling certain characteristics. The researchers interbred more than two hundred tomato plants for three traits: fruit mass, concentration of soluble solids, and pH, or acidity. They tracked the inheritance patterns of these traits and of more than three hundred genetic markers to figure out where the genes controlling the three traits are located.

Lander succeeded by combining Mendel's lessons about heredity, DNA analysis, and sophisticated math. The result was the first-ever discovery of how specific genomic regions were linked to complex traits involving more than one gene—and proof that this sort of analysis was possible.

With Lander's help and methods, Jacob turned to the work he'd discussed early on—mapping the genes controlling blood pressure in rats. For Jacob, this involved hours of "incredibly boring" work that consisted of randomly sequencing rat DNA to identify variations in repeats of the letters A and C that could serve as markers. While Jacob found the markers, others in the lab worked on a software program to sort through the wealth of genetic information he was discovering. Ultimately, the lab created a genetic linkage map that used 120 markers across the genomes of more than a hundred rats bred from parents with high blood pressure and parents with normal blood pressure. At a time when computational analysis was in its infancy, the scientists generated tens of thousands of data points for their software to analyze. The effort paid off. Once they crunched their data, they found strong evidence that one particular gene had a major effect on blood pressure in rats.

It was a significant step, and Jacob and Lander raced to publish. They knew that another research team was closing in on a similar finding. The two stayed up all night in Lander's basement, putting the finishing touches on their paper. Lander sat at the computer typing, Jacob at his side reviewing data.

"We did not sleep, and when we emerged the next morning, we had a fully written paper that we then took over to the journal *Cell*," Lander recalled.

Thirteen days later, the prestigious journal published their article. Francis Collins, a leader in the effort to sequence the human genome who would later head the National Institutes of Health, was a reviewer. Jacob was the first author on the paper. This was world-class science, and Jacob was at the forefront.

The young researcher went from knowing relatively little about genomics to breaking new ground, eventually building the first rat genetic map "and then going forward and really driving rat genomics in a big way," Lander said.

Jacob was on track for a big-league research career. He became an assistant professor at Harvard Medical School, where he ran his own lab. When Cowley came calling in 1993, Jacob had more than fifteen people working for him. They were creating some of the first genetic maps for the rat and the zebrafish, another organism used to model human disease.

Still, Cowley believed he could offer Jacob a better track. He knew the young scientist had identified about four hundred markers—and he reasoned that until the entire DNA code was cracked, those markers were a gold mine that could contribute to the discovery of all of the genes responsible for hypertension. If they could find the genetic basis of hypertension, they would gain valuable insight into heart attacks, strokes, and other conditions that result from abnormally high blood pressure.

Such an effort would require hundreds of markers, the mathematical techniques developed in Lander's lab, and the ability to breed generations of rats—all of which Jacob had. And Cowley could provide something many other research organizations lacked, the physiologists necessary to make it all happen.

Lander clearly understood the process that Cowley was envisioning. In a chapter he wrote for a book published in 1991 called *Very Large Scale Computation in the 21st Century*, he described how it would be possible to find the approximate locations of genes linked to high and low blood pressure. Researchers could breed rats with those characteristics, mate them, and trace markers that would follow those traits through different generations. Because they knew the locations of the markers, they could use Lander's mapping technique to locate regions that contained genes for the traits. But Lander was not likely to pursue such research; he was too deeply involved in efforts to sequence both the human and mouse genomes.

Cowley was not certain he could convince Jacob to leave Boston for a small medical college the Harvard researcher had barely heard of. But he kept trying. The conversation that began that afternoon in 1993 on the beach in La Jolla continued for nearly three years as Cowley patiently courted Jacob. Again and again Cowley made his point: linking genetic data to bodily function required physiological expertise—Cowley's expertise.

What was Jacob going to do with all of those markers? He would need physiologists to help him breed the lab animals and study their traits. But as molecular biology took hold, physiology departments were shrinking, and Harvard's was no exception.

"How can you remain in Boston? They've killed the physiology department there. There's no one for you to work with," Cowley insisted.

He sent Richard Roman, one of his physiology professors, to Boston to learn genetic techniques from Jacob and to continue the recruiting effort. Roman returned pessimistic. Jacob had been identified as a superstar. He had drug company funding and all of Lander's equipment at his disposal. Plus, he might have a case of "Harvard-itis," Roman reported.

The Harvard/MIT cluster had nurtured many Nobel Prize laureates, including James Watson, Walter Gilbert, Har Gobind Khorana, David Baltimore, and Salvador Luria. Boston was only two hundred–some miles from Cold Spring Harbor Laboratories, where Watson now presided over one of the nation's preeminent centers for genetics research. It was right where a rising genomics researcher would want to be.

But Cowley knew Jacob well enough by then. If he could secure the funding, he believed, he could get the Harvard researcher to come to the Midwest.

Six months after Roman's visit to Boston, he saw Jacob again at the American Society of Hematology annual meeting in New York. They took their wives to dinner at Tavern on the Green. Afterward, when the couples stepped out into the chilly spring night, Roman suggested a carriage ride through Central Park. Streetlamps glimmered. Apartment buildings towered above them. The horses trotted past the Ritz-Carlton.

"We gave it a good shot, but you're not going to come, are you?" Roman asked.

"Milwaukee was not among the top two hundred places I would choose to live," Jacob replied. "But try me."

This time Roman was encouraged.

The decision, of course, was not Jacob's alone. It helped that his wife, Lisa, liked the idea from the beginning. An Iowa native, she welcomed a return to the Midwest. "The best thing about Harvard is not being there," she said. "It's being *from* there." Neither of the Jacobs would miss the East Coast arrogance.

Lisa had come to think that her husband could be like a helium balloon: it was her job to tug on the string every so often, to ground his breathtakingly grand plans. Cowley's proposal had gotten Jacob thinking. He was looking at other jobs. Unbeknownst to Cowley, Jacob had just accepted an offer to be chief scientific officer at a Huntsville, Alabama, company called Research Genetics. But Lisa thought Cowley's offer made more sense. A move to Milwaukee would get them back to the Midwest just as they were finally starting a family.

Cowley sent a letter telling Jacob that since he hadn't heard from him, he would begin looking for someone else. That got Jacob's attention. Normally very logical about his career moves, he realized that something didn't feel right. Two weeks after accepting the offer in Alabama, Jacob told Research Genetics he had changed his mind—he would be taking the job in Wisconsin.

Jacob knew he could have an impact at the Medical College of Wisconsin. But he would be taking a big risk. Dzau told him that this could be the biggest career mistake he would ever make. After all, he was giving up a coveted position at a prestigious institution that was already deeply involved in the new science that he was pursuing.

Cowley, too, had much at stake. He had convinced the dean to find the money to bring Jacob on board. The Medical College would provide about $1 million, enough to renovate a 3,000-square-foot lab, stock it with equipment, and bring several of his team members to Milwaukee. Cowley was placing a huge bet, the largest the school had ever made on a junior faculty member.

"My head was probably the biggest head on the block," he later said. "If this didn't work, it would be Cowley's folly."

Jacob enjoyed the electric conversations, the promise of collaborating across departments. Cowley had the right pedigree, the right vision. He had good physiologists who would work with Jacob on better understanding the brain, lungs, heart, and other organs. And he had a facility with the rats needed for experiments, about 1,500 of them.

In Boston, the researchers and clinicians were on different campuses, about a twenty-minute bus ride apart. But in Milwaukee, when Jacob walked the hallways, rode the elevator, or ate lunch in the cafeteria, he would be bumping into doctors who were treating patients. It would be a constant reminder of the point behind all of the rat research: improving human health.

When Jacob finally announced his intention to leave Massachusetts General to take the Medical College's offer, John Potts, the hospital's physician in chief and director of research, wanted to know why. Jacob still had doubts about the move when he entered the boss's office to explain; Potts would erase them.

Years later, Potts said through an assistant that he had no firm recollection of Jacob or the conversation that would change the young scientist's life. But Jacob never forgot the question Potts posed.

"So let me understand this," he recalled Potts saying. "You're leaving man's greatest hospital for the Medical College of Where?"

Chapter 4

The Pied Piper

1996–2004

From an early age, Howard Jacob was driven by two compulsions: a need to explore the unknown and a desire to lead. At the Medical College of Wisconsin, he was able to satisfy both.

Jacob arrived at the school in 1996 with eight employees and $900,000 of funding from federal grants and Bristol-Myers Squibb. His new colleagues soon realized the extent of Jacob's ambitions.

"Howard thought big, and he allowed the rest of us to think big with him. He taught us to be opportunists," said Andrew Greene, a Medical College physiology professor who had trained as a biomedical engineer.

While still in Boston, Jacob had begun working with Allen Cowley on a five-year, $6.9 million federal grant to establish a Specialized Center of Hypertension Research. This project was vastly more comprehensive than the typical single-gene studies of the time. The two men pulled in other researchers, and using Jacob's rat markers, along with others that had been identified for humans, set out to find all of the genes affecting hypertension. It was one of the first efforts anywhere to determine the genetic basis of a complex disease.

The work would result in a paper, published in 2001 in *Science*, that provided the first genomic maps of the rat's cardiovascular function. They would publish another paper a year later in the journal *Hypertension* that connected characteristics such as body mass with certain chromosomes in groups of African American subjects. A third paper in 2005, in the *American Journal of Human Genetics*, identified areas on particular chromosomes that correlated with heart disease in humans. These were big leaps, all in quick succession.

"Howard and I, in those days, just kind of lived together and dreamed together," Cowley said.

They organized a conference, sponsored by the American Physiological Society, of which Cowley was president. Held in New York state at Cold Spring Harbor Laboratory, the home of modern genetics, their 1997 Banbury Conference brought together a hand-selected group of thirty-three scientists from the fields of molecular genetics, physiology, and pharmacology. James Watson opened the meeting. With the Human Genome Project still under way, the scientists developed five goals for a "Genes to Health Initiative," a road map for combining functional biology with the rising tide of genetic knowledge.

Jacob received tenure that same year and did something Cowley had predicted. He started a company. PhysioGenix bred genetically diverse rats for research use.

In 1998, when Cowley began a four-year term on the advisory council of the National Heart, Lung, and Blood Institute, part of the National Institutes of Health, he pushed the agency to require that all of the hypertension research it funded involve genetics.

They were redefining physiology. It was becoming physiological genomics.

In 1999, Jacob was named to head the Medical College's new Human and Molecular Genetics Center. He now could advance knowledge of the genome even further. The center's goal was deceptively simple: to translate genomic sequencing from research into improved health care. But this would require more than just a better understanding of the genetic nature of many common diseases; it would mean integrating that knowledge into the complicated practice of medicine. It was a daunting goal, but genetics had been heading for some time toward the day when knowledge of the code would translate into improvements in human health.

Lisa Jacob had no doubt her husband was well equipped for the challenge.

"Whatever Howard does, he's not going to do it a little bit, it's going to be overdone," Lisa said. "He's going to be pushing and shoving as hard as he can."

In 2000, Peter Tonellato, a Moroccan-born mathematician who worked closely with Jacob, won a five-year, $7.3 million grant from the National Heart, Lung, and Blood Institute to establish a rat genome database that would collect information on the lab animal from researchers around the world. The project hired more than a dozen employees and published its own quarterly newsletter, *The Pied Piper*, a nod to the rat-related medieval tale. The newsletter highlighted new discoveries in rat genomics and physiology and was distributed to thousands of researchers around the world.

That same year, 2000, Jacob helped the physiology department get its largest-ever grant under a National Heart, Lung, and Blood Institute program aimed at developing tools to link genes to biological function. The grant would run for ten years and bring in more than $25 million for studies of physiological and genomic stressors in inbred rats. The Medical College was the only research center in the Midwest to be funded through the program.

The school was becoming one of the biggest rat centers in the world. By 2001, the number of lab rats housed there had nearly quadrupled to 5,800, from about 1,500 when Jacob arrived. There would eventually be more than four hundred strains, including rats bred to have alcoholism, diabetes, hypertension, bleeding disorders, and other conditions; researchers used them for tests they couldn't do on human subjects.

By 2004, the Medical College was among fifty-eight institutions led by Baylor College of Medicine's genome-sequencing center that published both the genomic sequence and the map of the rat genome in the journal *Nature*. Jacob selected the strain to be sequenced, participated in many of the high-level meetings, provided the inbred rats and some of the maps needed to assemble the genome, and worked on the database to make the information useful.

The geneticist, however, continued to view his work with the rats as a step toward the goal of improving human health.

Historically, there was a wall between researchers and clinicians— and Milwaukee was no exception. As part of the package he negotiated with Cowley, Jacob had received the title of associate chief of genetics in pediatrics at the nearby Children's Hospital of Wisconsin. But he did not always get the access he wanted. He and Robert Kliegman, the hospital's pediatrician in chief, had a rocky relationship at first, dominated by turf battles.

Kliegman, who wore the yarmulke and long beard of an Orthodox Jew, was acutely aware of how children with rare diseases often went undiagnosed unless they had what he called the "geographic luck" to land in the right specialist's office. Inspired by work he had done during medical school in a Columbia University cancer research lab, he had his own ideas about bringing genetics into the clinic.

But a lucrative offer Jacob received in 2002 from two Ohio-based institutions had the unintended consequence of bringing the two men closer together.

Cleveland Clinic and Case Western Reserve University had

joined forces to offer Jacob a job that would provide him with $100 million to build a program in personalized medicine. Bringing genetics into the clinic to help patients was exactly what he hoped to do. However, despite assurances from senior leaders at both organizations that they were behind the new center, he was cautious. "Everybody in the trenches in both institutions told me that these institutions had been fighting with each other for years and it was never going to work," Jacob said.

Michael Dunne, the dean of the Medical College, asked Kliegman to speak with Jacob to try to stop him from accepting the offer. Understanding the importance of keeping Jacob in Milwaukee, Kliegman did not mince words.

"We sort of shook hands, and I said, 'I'll open the door to pediatrics for you,'" Kliegman recalled.

Access to real patients was what Jacob wanted most. He decided to stay.

By the time the rat genome was sequenced in 2004, Jacob could see very clearly the trajectory genomics would follow. It took more than ten years and about $600 million to sequence the first human genome. It took only about three years and just $100 million to sequence the equally complex rat genome.

"At that point I was convinced the technology curve would continue to bend," Jacob said. "I thought in ten years, this will be cheap."

In fact, the goal of sequencing an entire human genome for $1,000 had emerged in the first year of the new century, even before the Human Genome Project was completed. It gained momentum in 2002 at a new technologies symposium hosted by Craig Venter, the renegade scientist who had mounted his own human genome–sequencing effort to compete against the government-funded project. The $1,000 target made sense to many researchers and doctors, because it would mean that sequencing a whole human genome would then be cheaper than some single-gene tests.

Jacob saw that sequencing was moving closer to the realm of medical care. He knew the Medical College didn't have the resources to compete directly with bigger players that had worked on the Human Genome Project. But he and Cowley were driving research that was giving them and their colleagues a better understanding of the relationship between the genome and human disease. And they had an edge: access to patients at nearby Children's Hospital.

In 2004, Jacob presented a bold timeline to the school's executive committee. They would begin sequencing patients in 2014.

The plan was ambitious but still left them time to prepare, to buy the right equipment, to hire the right people, and to develop systems to evaluate and select cases for sequencing. That way, when the right child with the right medical predicament arrived at the hospital, they would be ready.

Chapter 5

An Extraordinary Patient

FALL 2004–EARLY 2007

Not far from the Medical College and Jacob's quest, in the small Madison suburb of Monona, a medical mystery is playing out inside the gut of a small boy.

It begins innocently enough one fall morning in 2006, when a mother goes to dress her son in swim trunks at a water park and notices a patch of red, swollen skin on his bottom.

The mother, Amylynne Santiago Volker, has four children: Mariah and Kristen, who became her daughters when she married their father, Sean, an electrician; Leilani, her daughter from a previous relationship; and Nic, her son with Sean.

Sean and Amylynne have been married for several years. They met through friends during a barroom pool game. He made a solid impression: a quiet, blue-jeans and pickup-truck kind of guy. A man's man in a state where masculinity revolves around deer hunting, ice fishing, and tailgating before Green Bay Packers games.

She liked the way he listened to her. Sean was humble, honest, straightforward, and a good father to his girls.

Sean had grown up in Monona, the youngest of four children. He was in some ways like an only child though, a full ten years younger than the next-oldest sibling. His father was a superintendent at Madison Gas and Electric. His mother, whose parents had immigrated from Germany, stayed at home until Sean started school. Then she got a job as the lunch lady, moving to each new school her son attended.

Sean earned decent grades but disliked classes, preferring baseball, football, and hockey. In high school, he was an all-conference lineman and cocaptain of the football team. Colleges recruited him, but Sean had no interest. He went to technical school for an associate's degree in electronics and an apprenticeship program that prepared him to work in construction. He liked using his hands and preferred the straightforward work of line jobs to the responsibilities that came with supervising. Amylynne noticed that Sean was meticulous—never the fastest but always the hardest worker.

While Sean's mother had stayed nearby, even during school, Amylynne's mother had been quite the opposite. Through much of her childhood, Amylynne lacked stability. Her father and mother had an unconventional and somewhat confusing relationship. He was a handsome doctor from the Philippines; she, a glamorous French American from the Chicago suburbs. They enjoyed ballroom dancing and belonged to the Playboy Club in Lake Geneva, yet for years they did not live together in the same home. Were they married? Divorced? Dating? Their marital status was kept secret, even from relatives.

In early adulthood, Amylynne embarked on her own adventurous and unpredictable path, trading the Midwest for Hawaii. Beautiful and full figured, she modeled swimwear and even took a job as Miss Smirnoff, the hostess at high-profile parties and yachting excursions. She stayed in Hawaii for nine years, starting her own business and living in a high-rise condo in a gated community with multiple swimming pools. But in time Amylynne felt the tug of home. After learning that her mother was homeless and that a neighbor who had helped to raise Amylynne was in failing health, she returned to Wisconsin.

When Amylynne met Sean, both were finished with their adventures in singledom and seeking a chance to settle down. She had a daughter, Leilani, from a failed relationship. He had two girls, Kristen and Mariah, from a failed relationship of his own. Both Sean and Amylynne had become born-again Christians.

Within three months of meeting, the couple had received the required counseling and permission from their pastors to marry. Amylynne, a very different woman from the "warrior mom" she would become, hated her wedding dress; it was puffy and lacy and not at all sophisticated. But her best friend had pushed her to buy it. Back then, she was not one to argue.

They held a private wedding at West Madison Bible Church. Already the new family felt comfortable together.

"I've seen longer kisses than that," Leilani, then six, yelled out during the ceremony.

The Volkers took the three girls to the nondenominational Mad City Church in Madison. They attended family Bible study. They volunteered at a local food pantry. They saw themselves as people of faith.

Early in 2004, Amylynne got pregnant. Sean hoped for a boy, though hope was an understatement. He saw himself going to baseball games with his son and tossing a football back and forth in the yard. The boy would be a Packers fan, a dominant trait in Wisconsin, passed from each generation to the next. Sean's heart was so set on a boy that he would not allow Amylynne to pick names for a girl.

Amylynne was not so particular. She had experienced two miscarriages. Either a boy or a girl would be fine, just not another miscarriage.

The pregnancy was difficult. Sometimes it seemed to Amylynne that she was struggling with nausea morning, noon, and night. She actually lost weight, at one point close to twenty pounds.

But on October 26, 2004, the Volkers got their wish. After seven hours of labor, the baby was born, a boy, 6 pounds, 13 ounces, and 19.5 inches long. Amylynne refused all drugs. Sean, who has difficulty with blood, could not cut the umbilical cord. Leilani, then eight, performed the task. An hour later, Amylynne showered and put on her makeup. She could not believe how easy the birth had been, even the labor.

The Volkers celebrated, giving their son a name as unique as his genes: Nicholas Zane Fernando Santiago Volker. Sean and Amylynne particularly liked the name Nicholas, the patron saint of children.

Nic, as the family took to calling him, appeared healthy from the start, though he had his own quirks. His quiet demeanor ended abruptly whenever Mama wandered out of sight. He wailed and would not be held or consoled by anyone but Amylynne. He didn't like sleeping in his crib or in his room. He crawled oddly, on one leg instead of two. Amylynne would drop to her hands and knees to demonstrate proper crawling technique. Even so, Nic showed little improvement.

Walking was another story. Nic learned quickly and was soon much better at walking than crawling. He was the boy his father had imagined, kicking and throwing balls in the yard of their Monona home. He ran to play on the swing set. He adored the family dog, a bichon frise named Kyler. He liked clambering onto the front seat of the family van and pretending to drive. The quiet little fellow evolved into a chatterbox.

Nic enjoyed playing with his older sisters but picked favorites. For a long time only Leilani could hold him. Then, for a while, it was Mariah. The one unbending rule was that no matter whom he played with, Mama had to be there, too. He was Mama's little boy.

The only cause for worry was the child's appetite. He had a certain pickiness when it came to food. Though he devoured Amylynne's breast milk, all other foods seemed to repulse him. His head, topped by a generous mane of light blond hair, turned away from the spoon. He lagged on the growth charts. It was a consistent problem, though not an urgent one. Amylynne tried various diets: carbohydrates, gluten-free, lactose-intolerant, foods for people with celiac disease. Nothing worked. Nic was still breast-feeding approaching his second birthday. Doctors told Amylynne she wasn't feeding him enough.

It went unspoken, but Amylynne felt the accusation just the same: *Bad mother!* Still, Amylynne was not one of those moms who took the physician's word as gospel. Despite, or perhaps because of the half-dozen or so doctors in her family, she had come to trust her instincts. Besides, Nic seemed normal enough on the whole. What child did not have some food issues?

But everything changes that morning at the water park in September 2006. The angry red welt on Nic's bottom is the size of a golf ball. Amylynne notices a discomfort in his walk. It is painful to watch.

Amylynne's first instinct is to take him to the emergency room. Instead she calls her father, the doctor. His advice: take the boy to the hospital and have doctors lance what is probably just an abscess.

Sean takes him. Father and son arrive at the University of Wisconsin Children's Hospital in Madison, where a doctor looks Nic over, gives him antibiotics, and says the abscess should go away. But that is not what happens. The abscess bursts soon afterward, and in its place, two holes appear at the opening of Nic's rectum. Now Amylynne is worried. Her own instincts, both maternal and medical, tell her something is very wrong with Nic. She is certain of it.

The wound from the abscess never heals. It's a gruesome ailment, difficult to witness, difficult to talk about, all the more difficult for Nic to endure. Soon, the two holes in his rectum merge, forming a much larger opening, which begins leaking stool. The little chatterbox, now two, falls silent. His appetite—such as it is—dwindles even further. His face contorts with pain.

Amylynne calls the hospital to schedule another visit. This time, after looking Nic over, the doctor says, "He has to be admitted right now."

It's in this moment that the parents realize their boy has no routine childhood illness. Nic is referred to a gastroenterologist, a specialist in ailments of the stomach, intestine, and liver. The parents now, for the first time, hear a name attached to their son's illness: Crohn's disease.

From the medical web sites proliferating on the Internet they learn that Crohn's is a form of inflammatory bowel disease. The illness is an autoimmune disorder, one of those curious conditions in which the body's defenses mistakenly attack healthy tissue. One hallmark of Crohn's is swelling of the gastrointestinal tract.

But there are two nagging problems with the diagnosis. Crohn's very seldom strikes children as young as Nic. Worse, Nic's illness does not behave like Crohn's. There are treatments for Crohn's, including medicines that reduce inflammation or suppress the immune system. Nic does not respond to the medicines. Nor does his condition improve when doctors try other treatments, including special diets.

Amylynne had felt devastated by the Crohn's diagnosis. Yet when the evidence suggests that Nic has something else, something not readily recognizable, she feels much worse. She is unprepared for uncertainty.

Amylynne likes Nic's main doctor at UW Children's Hospital, Robert Judd, a pediatric gastroenterologist. He is a nice man, caring. But like other old-school doctors she's met, he isn't especially communicative. He tells her little, though perhaps it is just that he does not know what to say in the face of his own uncertainty. The medical staff run numerous tests. They take blood. They check for cystic fibrosis. They search inside the lining of Nic's intestine with a lighted tube: a colonoscopy, a common procedure in adulthood but not in childhood.

Early on, some doctors talk of a rare metabolic disease that can affect the intestines: CDG, or congenital disorder of glycosylation. But none of the theories leads anywhere. Amylynne feels she is not receiving enough information from the medical staff. She finds it difficult even to get test results.

Nic's illness, first discovered at the water park in October, lingers and by the end of 2006, the ten-mile drive to the hospital grows more routine and more frustrating. For one thing, Nic has lost weight.

"He looks like he's from the Holocaust," one physician tells her. But when she tries to feed the boy cereal and other foods, he eats little. No one seems to know why he is sick.

As far as Amylynne can tell, the staff is not following any particular plan during Nic's hospital stays. A nurse will enter the room and announce to the mother and son, "You're leaving." Minutes later, the doctor pokes his head around the door: "You can't leave today." They're in the hospital, then back home, then back in the hospital. Nic takes tests and more tests, sometimes the same test again and again.

Months pass, and Amylynne fights a rising sense of panic. She is always waiting for a test result or struggling to get paperwork from the hospital. She has asked for a summary of Nic's care to take to Children's Hospital of Philadelphia, well known for its work in gastrointestinal medicine. But the summary is never ready.

Augmentin, the antibiotic that doctors have prescribed, appears to do little. The feeding tube they thread through Nic's nose only annoys him. He is constantly yanking it out.

One test the child receives, a fistulagram, searches for holes called fistulas in Nic's intestines. The test involves sticking a very long needle into the boy's bottom. Nic is too young to receive drugs for the pain, and his cries are awful; they tear at his mother. The doctors find no fistulas.

Months spent in the Madison hospital leave the Volkers certain of two things: Nic is not getting better; and despite an endless number of tests, the doctors are no closer to understanding what's making him sick.

The mother can see the pain on her son's face. He is so small, so shy, so clingy. He is helpless to explain what is hurting him. He'd been progressing so well with his talking. But by January 2007, three months after the emergence of his symptoms, his talking stops.

While he had once been comfortable breast-feeding, now he struggles. He clings to Amylynne as if she were a pacifier, but he does not eat. And he keeps losing weight. The lost weight and Amylynne's continued breast-feeding fuel suspicions among some doctors that she might have something to do with her son's illness. When they ask about his diet, Amylynne senses an accusatory tone. She and Sean understand that behind the doctors' questions is a search for blame, and it hurts them.

Most of all, though, the Volkers share the helplessness felt by the medical staff. Every time Amylynne sees the hole in her son's bottom, she cries. He has such a cute butt, she tells herself. Now it's going to have a scar. She knows, of course, that there are worse things than scars.

She cries when the medical staff inserts the feeding tube. She cries through doctor visits. When the nurse case manager needs to schedule a meeting at the Volker home to insert a tube through Nic's nose for feeding and medication, she picks a time when Sean will be there. "I know you can't handle it," she tells the mother.

With each new development, Amylynne's frustration grows. Sometimes she wants to scream: What's wrong with my son?

She and Sean believe God will protect Nic, but his illness forces them to consider the worst possibilities. What if doctors never figure out his disease? What if Nic dies before he can choose to be baptized? Their religion tells them to have faith, but the confusion and helplessness of the hospital tells them to prepare. They ask their pastors to come to the house to dedicate Nic. In this ceremony, an infant is committed to God without being immersed in water.

Amylynne takes on more and more tasks associated with her son's care. Once, however, she asks Sean to take Nic to the hospital to get the feeding tube reinserted. The bill tops $1,400. That is enough to change Amylynne, a woman who had once been squeamish to the point of throwing up at the sight of excrement. She learns to insert the feeding tube herself. Nic's illness does not allow her the luxury of being squeamish; she must get over it. The disease keeps pushing her out of her comfort zone, testing her strength and patience.

One day she takes Nic to the hospital for an appointment and they wait three hours in a tiny exam room before the doctor sees them. Amylynne smolders. She has reached her limit. She decides she is finished with the hospital in Madison.

Amylynne seeks a second opinion. She wants to go to the Cincinnati Children's Hospital Medical Center, but doctors in the family suggest she stay closer to home. They recommend a young gastroenterologist named Subra Kugathasan. Kugathasan works at Children's Hospital of Wisconsin near Milwaukee, a little more than an hour from the Volkers' home.

Children's Hospital of Wisconsin ranks among the best pediatric facilities in the country and is among the busiest, treating some sixty thousand patients a year in its emergency department and trauma center. Just a few years before Nic's arrival, the hospital made headlines for its groundbreaking treatment of Jeanna Giese, who, in 2004, became the first person in the world to survive rabies without vaccine. Children's shares a campus with an adult facility and the Medical College of Wisconsin. Many of the doctors at

Children's have joint appointments with the Medical College, including Howard Jacob and members of his genetics team.

Kugathasan has treated hundreds of patients with inflammatory bowel diseases, illnesses that appear similar to Nic's. He is something of an expert in the field. Yet he has never seen a case quite like Nic's. The severity of the disease is remarkable. Nic arrives at Children's in May 2007, a two-and-a-half-year-old boy weighing just seventeen pounds: that puts him below the third percentile on the growth charts; average for a boy his age is about thirty pounds.

Like many of the doctors who have seen Nic and will see him in months to come, Kugathasan examines the symptoms and finds it hard not to think of Crohn's disease. He puts Nic on a Crohn's medication called Remicade that reduces fistulas but also weakens the immune system. Like the other doctors, though, Kugathasan runs into problems with the Crohn's diagnosis. The boy is simply too young, the illness too resistant to treatment.

Looking back, Kugathasan would put it this way: "He basically failed every single medical, surgical, and nutritional approach."

Still, Amylynne likes Kugathasan. He is gentle, thoughtful, nonjudgmental. He is well dressed, which might seem an odd thing to be impressed by, but it reminds her of her father. She trusts Kugathasan.

She grows fondest, though, of Nic's surgeon, Marjorie Arca. Like Amylynne, Arca has a Filipino background. She grew up in a rural area of the Philippines before moving to the United States with her family at the age of twelve. She'd wanted to be a doctor, but for years, as she went through the different medical training programs, she could never decide on a specialty. Then came the moment she would always remember, the moment when she chose pediatrics, or rather, when pediatrics chose her.

There was a four-year-old boy named Zack, one of numerous children she had been caring for at C. S. Mott Children's Hospital at the University of Michigan. Arca, in residency for general surgery, had been up all night and was tired. Her feet were dragging, but she heard Zack had been looking for her. She steeled herself, not knowing what to expect. When adult patients want to see the doctor, often it's because they have a problem or complaint. Zack was different. The young boy had been looking for Arca to give her a present, a Christmas ornament he'd made. The resident's face eased into a smile. She would have trouble putting the feeling into words until years later when she had children of her own.

"To have somebody trust you with their child's life," Arca said, "that should make you strive to be your very best."

She feels that deep trust when she meets Nic and his mom. It's clear to her how much Amylynne loves her son. Arca sees the look of worry on Amylynne's face and understands what it means. It's a look that says to Arca "This is my precious child. This is the child I would die for in a heartbeat, without question."

Chapter 6

Diagnostic Odyssey

MAY–SEPTEMBER 2007

On her first examination of Nic, Marjorie Arca discovers four fistulas, holes the size of pencil pricks, running from the child's intestine all the way to the surface of the skin. Just weeks earlier, the doctors in Madison had examined the boy and found no fistulas. In addition to the holes, Arca finds two deep tears in Nic's rectum. But it is the fistulas that worry her most. Stool leaks through them onto a wound on his abdomen. Left unchecked, the situation could trigger a life-threatening infection.

To head off the threat, Arca performs a colostomy, bringing one end of Nic's large intestine out through the abdominal wall so that his waste can be diverted into a bag resting against his stomach. She hopes the procedure will preserve the use of Nic's intestinal tract and give his bottom a chance to heal.

The disease, however, has a will of its own. Sometimes when doctors adopt a new treatment, Nic's condition improves for a week or two. Then the illness reasserts control. The number of fistulas grows, and so does the puzzle.

The only certainty is that the disease isn't Crohn's. So what is it?

If the Volkers were unprepared to learn their son had a serious illness, they were even less prepared to hear "We don't know what it is." Parents go to the hospital expecting to learn what's wrong. Yet from time to time doctors encounter diseases that appear nowhere in the medical journals. Families unlucky enough to have such an illness are often forced into long, frustrating searches, going from doctor to doctor, hospital to hospital, research center to research center. These journeys are known as diagnostic odysseys. How many wind up on these odysseys isn't known, but government estimates offer a clue. Between 25 million and 30 million Americans have a rare disease, roughly one in ten. About 30 percent of them go five years or longer before receiving a diagnosis for their illness—almost 10 million people.

In May 2008, one year after Nic's arrival at Children's Hospital, the National Institutes of Health will establish the Undiagnosed Diseases Program. Using the latest technologies, including machines that read the genetic script, the program will seek to diagnose patients with mysterious conditions. In 2007, however, the doctors treating Nic are largely on their own against a disease that appears new and relentless.

Various specialists, experts on different systems in the body, join the hunt for a diagnosis. One is William Grossman, whose field is immunology, which focuses on the body's network of defenses against infections and other invaders. As early as third grade, the Minnesota native began listing "physician-scientist" whenever he was asked what he wanted to be when he grew up. By the time he entered high school, the AIDS epidemic was reaching its peak and Grossman found the challenge of it compelling. The immune system had become one of the most exciting and important areas of medical research; so much lay undiscovered.

He received his doctorate in immunology from Washington University School of Medicine in St. Louis, then in 2004 took a job at Children's Hospital of Wisconsin. There he quickly developed a reputation for taking on the hardest cases, the illnesses that baffled and frustrated other doctors, as AIDS had. Immunology offered the ideal perch from which to treat these children. For many doctors, the immune system is a black box. At Children's, a joke circulated that Grossman was running his own service for kids with unknown medical problems.

In 2007, the gastrointestinal specialist Subra Kugathasan brings Grossman onto Nic's team. The child is so sick that doctors are forced to consider two possibilities: either he is the victim of terrible neglect at home, or he has something "organically wrong with him," as one doctor puts it, something in his body's machinery that does not work.

Grossman searches for that flaw in Nic's immune system. He performs numerous tests, and his colleague James Verbsky will go on to perform many more, both of specific genes and of the immune system in general. What Grossman finds is troubling.

"He had some very abnormal immunological results," the doctor says. Nic has defects in what are called natural killer cells, which perform a critical role in the immune system. Responding to viruses and tumors, they live up to their name by attaching themselves to infected cells and releasing granules that destroy them. In Nic's case, the cells are failing to kill at a sufficient rate. Still, it is by no means clear that fixing the boy's natural killer cells will cure his disease.

Much about how the disease works remains mysterious, and attempts to find answers run into a chicken-and-egg question. Is the child's disease causing his nutritional problems, or are the nutritional problems causing the disease? The two are clearly related. Inflammation, the immune system's response to an attack, has nutritional consequences: it prevents us from absorbing nutrients, weakening the body and resulting in more infections.

Sorting out whether Nic's poor health is being driven by his disease or by poor nutrition "was one of those forks in the road," Grossman recalls. And it's connected to a related dilemma: Nic is malnourished, in desperate need of food; yet when he eats, the disease roars to life.

Nic's condition improves when he receives total parenteral nutrition—glucose, amino acids, vitamins, and minerals delivered through an intravenous line into a vein. Each time the doctors wean Nic off the intravenous feeding and allow him to consume real meals, the disease returns. Back come the painful holes in the intestines. The holes are so bad that one of the nurses cannot take the sight of Nic's bottom and asks to be removed from his case. Even doctors who specialize in inflammatory bowel disease have seen few cases as severe as Nic's.

The intravenous feedings are a temporary measure, funneling nutrition directly into his veins and bypassing the inflamed digestive system. But they are not a solution. They merely sustain him while Grossman and others search the medical literature and run their tests.

The colostomy that diverts Nic's waste into a plastic bag helps—but only for a short time. Less than a month after the surgery, more fistulas form. The skin around the holes turns a sour shade of purple and his condition worsens. The surgical wound from the colostomy should be healing, but it isn't. Instead stool is leaking from his intestine and threatening to infect the wound, the situation doctors have feared all along.

By late June 2007, several months before his third birthday, Nic is fading fast. Kugathasan tells Amylynne that they have underestimated the boy's illness; it is much worse than they thought. She can see that for herself. The pain is wearing Nic down, consuming all his energy.

In the last week of June, a friend records a hopeful verse from Galatians in Amylynne's online journal: "Let us not become weary in doing good, for at the proper time we will reap a harvest if we do not give up."

The words appear prophetic. Doctors give Nic blood transfusions and a medication to spur the production of infection-fighting white blood cells. The child fights through fevers. He goes days without solid food. He endures more blood transfusions. The number of his trips to the operating room reaches nine or ten—Amylynne loses count. The medical bills rise ever higher. In the next few years, Nic will exceed his $2 million in lifetime insurance benefits.

Slowly, however, the growing armies of white blood cells help Nic rally. His energy returns. He becomes well enough to play with his trains. His cheeks fatten for once. His weight creeps up to 24 pounds. It is still much too low for a boy approaching his third birthday, but he is well enough to play with his sisters and even leave the hospital for a few days at a time. Best of all, the fistulas begin to heal.

On September 1, Nic tells his mom he is not sick anymore. Nic's doctors concur: the child has improved. They decide that he is well enough to go home.

Chapter 7

Spiders on the Ceiling

SEPTEMBER–OCTOBER 2007

Two days after Nic pronounces himself cured, the Volkers race along Interstate 94 from Madison, bound for the emergency room at Children's Hospital. The weekend got off to such a good start. The boy was home from the hospital, blood counts good, smiling, energetic, back to his playful self.

Then abruptly, his breathing turns rapid and shallow. He grows lethargic. Amylynne presses her hand against his forehead, and a wave of fear sweeps over her. He is burning up. Soon he is vomiting.

Amylynne has some idea of what is happening. Her family's experience in medicine has taught her to recognize the classic symptoms of sepsis, a massive, often fatal syndrome in which the immune system goes into overdrive to fight a blood infection. Nic has all of the hallmarks: the fever, the rapid breathing. They have to get him to the hospital quickly.

Sean drives. Amylynne sits in back, comforting Nic. The boy's pulse gallops, and he mumbles. What he says alarms her: he sees spiders on the ceiling.

While Nic hallucinates, his mother prays aloud; there have been enough crises in his short life that the words flow easily, almost mechanically.

"Oh Heavenly Father, I ask that you please protect Nic."

At Children's Hospital, doctors quickly confirm her fears. Nic is in septic shock, a life-threatening disruption of the circulatory system. Usually, the system sends more blood to the organs and tissues that need it most and less to the lower-priority regions. But when a patient goes into septic shock, this supply-and-demand logic crumbles. Blood carrying oxygen and nutrients is diverted away from the brain and heart, pouring instead into organs that need it less. Nic's hallucinations mean that regions of his brain are not getting enough blood flow. Ultimately, septic shock can lead to organ failure and death. A quarter to half of all septic shock patients do not survive.

Nic is admitted straight to the intensive care unit.

George Hoffman, the attending doctor in the intensive care unit, is not optimistic. He looks at Nic and he thinks: There is a strong likelihood this child will die.

After twenty years at the hospital, Hoffman has seen enough to know that even the worst situations are not hopeless. He and the other doctors in intensive care begin by bombarding Nic's infection with antibiotics—not just one or two designed for specific pathogens.

"You use a shotgun, not a rifle," he says.

The doctors do not know the specific agent making the boy so sick, and it will take between twenty-four and seventy-two hours to identify it. So they use many different antibiotics and hope the combination will work. They have only about six hours to turn the tide.

Treatment halts Nic's downward slide, but his recovery stretches well beyond those crucial early hours. For days that turn into weeks, Amylynne's journal catalogues an ever-evolving series of threats. Some nights doctors tell her they don't think Nic can survive. Sometimes he grows so sick that Amylynne will not leave his side. At times she has friends write the entries in her journal for her.

> *Sept. 4 . . . Nicky is in dire straits. . . . He is at risk for all organs being damaged or failing now. His temp is now at 106. . . . He is delirious, hallucinating and trying to pull out all his lines.*

The 106-degree temperature puts Nic frighteningly close to the edge. A temperature of 108 is almost always fatal. Doctors fight back, administering hormones that make the heart beat strongly and increase blood pressure.

Amylynne asks members of her church, seventy-some miles away, to drive to the hospital and lay hands on Nic. Time is of the essence, she writes.

She camps out in her son's room.

Sean works and cares for the girls at home. He cannot bring himself to admit the possibility of his son's death. Yet the threat hovers over the child and runs through each new report from the doctors who examine him.

> *Sept. 6 [written by one of Amylynne's surrogate authors] . . . The doctors are asking [Amylynne] to make decisions about whether or not to resuscitate him if he would go into cardiac arrest.*

Faith will not allow her to say "do not resuscitate." How does a mother tell someone to let her child die?

Septic shock patients typically wax and wane in intensive care for anywhere from one to four weeks. Nic battles the infection. But he also struggles with the discomfort of a collapsed lung.

Amylynne can only watch and pray. She has little control. But to the extent that she can, she makes her own rules. Some nights she curls up beside Nic in bed, though sleeping with your child is not permitted. The nurses see her there, but they let her be. They marvel at how she maintains hope, how she manages each morning to pull herself together. After another of those nights when the doctors tell her Nic might die, this mother somehow calms herself, exhales, washes her tear-streaked face, and dresses in a clean business suit.

Sept. 13 . . . Nicholas has a couple of breathing scares because of his collapsed lung. . . . He is alert now, which makes it really hard because he is scared and in pain.

Nic's condition passes from one emergency to another, and relatives try talking to Amylynne about visiting a funeral home. They try to prepare her for the worst—not to scare her but to steel her.

She will have none of it; she refuses to set foot inside a funeral home.

Sept. 20 . . . Nicky is in Day 5 of a very high fever which has reached as high as 105.6. He has tested positive for E. coli which has come from his breathing tube. He also has pneumonia.

Amylynne's father, the doctor, encourages her to give the "do not resuscitate" order.

"Absolutely not," she tells him. "I won't give up hope."

In her journal, there is a word she will not use, the word the professionals seem to utter so easily. She calls it "the D word." And each time a doctor or nurse uses the word, she feels increasing anger and resolve.

Somehow she is right. The word proves premature each time

another expert uses it. The powerful antibiotics help her son's body to clear the infection. Nic receives numerous transfusions of blood, plasma, and platelets. His blood flow improves.

The little boy, cursed with this strange disease, also seems to have been blessed with a remarkable capacity to fight through each new assault.

The longer patients stay in the hospital, the greater the chance that something will go wrong, especially if they're young, old, or struggling with a weakened immune system. Their care comes down to playing defense, anticipating threats, and guarding against each one.

When a new infection enters Nic's body through an intravenous line, doctors replace the line. Slowly the fevers begin to abate. Nic's temperature eases back down. He gains a little weight, passing 25½ pounds; though this still places him in only the third percentile. Even so, it is the first time in fifteen months he has shown any weight gain. In the dour hospital room, his personality appears in flashes. To avoid pokes and prods from the nurses, he plays possum whenever they walk in, pretending to be asleep. It is what passes for control in his world.

Sept. 27 . . . He is looking good and laughing and playing with his sisters as I write this. Yesterday I finally got to hold him in my arms again and I cuddled him for hours like he was my newborn . . .

By October, a month after he saw spiders on the car ceiling, Nic is well enough to eat scrambled eggs, the first food in all that time that has not come through a tube. After mending slowly, his surgical wound from the colostomy has finally healed. At month's end, seven weeks after the sprint to the emergency room, Nic goes home to Monona. He is there in time to celebrate his third birthday.

Amylynne has stood firm, defying talk of "do not resuscitate" orders and funeral homes, steeling herself against the fears of relatives and medical staff. And Nic has borne out her tenacity. If his disease lurks somewhere in his genes, perhaps his mother's strength does, too, buoying him up when he needs it most.

Chapter 8

Big Steps and Fast

NOVEMBER 2007–JANUARY 2008

From his office, Howard Jacob can see the roof where the emergency helicopters land, carrying the most desperately ill and injured patients bound for Children's Hospital. The helicopters arrive at all times, day and night, and they strengthen Jacob's appreciation of one priority above all others: urgency. The hospital is a place where people cannot wait for research and institutional planning to catch up with their needs. Plans take a backseat to patients.

Medical challenges of the highest order enter the emergency room without warning. Who would have predicted, for example, that in the fall of 2004 an ambulance would deliver a dying rabies patient named Jeanna Giese to Children's? And who would have guessed that, in a last-ditch effort, a doctor named Rodney Willoughby, Jr., would craft a unique treatment to save Giese, making her the first human in the world to survive rabies without a vaccine?

You just never know.

You could plan to start sequencing at the hospital in 2014. You could draft white papers. You could show them to deans and boards and advisory committees. You could follow your schedule to the letter. Then a patient with a devilish disease could show up in the ER and rewrite the plan.

By 2007, Jacob is a familiar presence in the halls of Children's Hospital. Bob Kliegman has made good on his promise to open the door to pediatrics, and Jacob has taken full advantage. After turning down the offer from Cleveland Clinic and Case Western Reserve, he has begun spending much more time at Children's Hospital.

"I think by committing to being here, he ramped it up," Kliegman later says. "Howard became very visible here."

Unlike other researchers with little hospital experience, Jacob has come to understand the routines, the decisions that have to be made, and the challenges posed by the very sickest patients. He has yet to meet Nic Volker or learn of his strange illness. But Jacob has had many conversations with doctors and residents involved in other complicated cases.

As associate chief of genetics in pediatrics, Jacob attends a meeting every Monday afternoon with all of that department's section heads. He is pulled in whenever Children's is recruiting for a genetics-related job, and the hospital's Children's Research Institute has funded some of his lab's work. His hires are taking on important roles. Ulrich Broeckel, Jacob's former postdoctoral fellow, has moved his research lab into the institute's building.

For the entire month of August 2004, Jacob even went so far as to don the white lab coat he generally shunned in order to accompany Kliegman on his 7:30 a.m. ward rounds. He listened as residents discussed what had happened the night before and shared his ideas about the most difficult cases. Jacob asked questions that made the residents think, and he saw the difference between making decisions in the clinic and in the lab.

Sometimes doctors had to make a life-or-death decision in the absence of data. Sometimes all of the obvious tests had been given already, and there was no time to go home to think it over. Decisions had to be made, based on education, experience, or maybe just a hunch.

"That was difficult to watch," Jacob said.

The more he has watched, however, the more convinced Jacob has become of the role DNA sequencing can play in medicine. It is a conviction he cannot shake, and it becomes the focus of some of the speeches he gives around Milwaukee. He tells the local chapter of the Young Presidents' Organization that genetics is going to improve people's lives. When he presents at grand rounds, a ritual of medical education in which interesting cases are described to other doctors, Jacob talks about how DNA technologies are poised to change the practice of medicine.

These are not yet topics Jacob broaches with a national audience. They are instincts and expectations, not the discoveries and facts that medical journals seek. Scientists judge one another by the work they publish, and Jacob's papers frame his role succinctly; he is the rat guy.

The rodents have loomed large in Jacob's scientific career, crucial to many of the 150-plus scientific papers he has authored or coauthored by the end of 2007. All of that work focused on finding genes more efficiently, says Richard Roman, his former colleague. Which, in turn, led to the ultimate goal.

"Everything we've done, for me has been about understanding human disease," Jacob says. "We built it from the standpoint that we could make discoveries in rats, solve the problem in rats, and export the solution into humans."

But in 2007, Jacob's efforts to make discoveries in rats were being slowed by a vexing issue: science's inability to manipulate their genes.

For some time, mice were the only lab animals whose genes

could be altered to create models for human disease. A trio of scientists—Mario Capecchi, Martin Evans, and Oliver Smithies—won the 2007 Nobel Prize in Physiology or Medicine for discoveries during the latter half of the twentieth century that allowed scientists to target and deactivate single genes in mice. The resulting animals, called knockout mice, opened the way for experiments that could examine the roles individual genes play in the body. Mice became the go-to animals for genetics experiments, even though rats are closer to humans in terms of some physiology, behavior, and cognition, and are the primary lab animal for testing new drugs. Mice were not ideal for answering some of the most important questions about gene function in humans. What geneticists lacked was a knockout rat.

Jacob hopes to change that by testing new technologies researchers are developing to modify rat genes. "Part of my job is to look out over the horizon," he says.

The first gene-editing tool he and his team experiment with is ENU mutagenesis, an agent that researchers have used successfully in mice and on a small scale in rats. It induces random mutations in sperm cells but is not well suited for making precise changes to specific regions of a gene, a necessary prerequisite for a knockout rat. Under the guidance of Aron Geurts, a postdoctoral fellow, Jacob's lab instead tries making edits using transposons, stretches of DNA known as jumping genes that can change position within the genome. Like ENU mutagenesis, this method disrupts genes, but it gives researchers more control in creating the mutations.

"But again it was far away from Howard's tastes because it was slow," says Josef Lazar, Jacob's right-hand man. "Howard is a guy going after big steps, not small steps. Big steps and fast."

Unlike Gregor Mendel, the Austrian monk who established many of the rules of heredity by cultivating some 28,000 pea plants over about eight years, patience is not Jacob's strong suit. His search for a faster gene-editing tool ends one day with a phone call from a stranger.

Fyodor Urnov, a senior scientist at Sangamo BioSciences, a California biopharmaceutical company, wants to show researchers the power of his company's new technology. Sangamo has developed synthetic proteins called zinc finger nucleases that the company believes can modify a cell's DNA at an exact location. All previous techniques have been clumsy and random—precision isn't their strength. Urnov has been told by Rudolf Jaenisch, a professor of biology at the Massachusetts Institute of Technology who played a pioneering role in the modification of mouse genes, that to prove the technology, he needs to show how it works in the rat. Jaenisch's reasoning is that the rat is the next step in genetics research: in terms of organ size, life cycle, and many other physiological characteristics, the rat is closer to humans than the mouse. This is a powerful affirmation, given that Jaenisch built his career on manipulating mouse genes.

After studying the field, Urnov has settled on Jacob, one of the best-known rat researchers in the world, to test the zinc finger technology.

"I was completely skeptical," Jacob says of Urnov's phone call. But he soon becomes convinced that the zinc finger could be a "game changer," allowing researchers to click on a gene and alter its DNA—as easy as inserting or deleting text in a Microsoft Word document. Being able to modify genes so quickly would hugely accelerate researchers' quest to learn how they work.

When Jacob tests them, zinc finger nucleases perform as well as Urnov has predicted. Injected into rat embryos, they quickly repair or inactivate selected genes. It is in Jacob's lab that the rat becomes the first mammal to be genetically engineered with zinc finger gene editing. The work becomes the basis of a paper that will be published in the journal *Science* in 2009 showing how researchers successfully eliminated gene functions and created knockout rats with specific genes deactivated. Zinc finger technology will ultimately help scientists manipulate genes to design rats with Parkinson's disease, Alzheimer's disease, hypertension, and diabetes in order to better understand those conditions.

What impresses Urnov most about Jacob, however, happens after the *Science* paper is published.

"What do I need to do to make this broadly useful?" Jacob asks.

He is willing to try untested technologies. Moreover, once they are verified, Jacob wants to make sure as many researchers as possible have access to them. Although the rest of the scientific world sees him as the rat guy, Jacob knows that the rats are really just another tool for better understanding the human genome.

To be sure, there have been other more publicized and more expensive efforts to understand human disease through DNA sequencing. A $100 million attempt to identify all of the common genetic mutations in the human population, called the International HapMap Project, had begun in 2002. The next key step, to compare patients' genomes with those of healthy people in what were called "genome-wide association studies," then began gaining steam. These GWAS (pronounced *G-waas*), as they were known, cost about $10 million each, on average. They used devices called SNP (pronounced *snip*) chips to scan portions of individual genomes, searching for places where there were single-letter differences. Typically, a study would compare the DNA of two groups: people with a given disease and people without it.

In June 2007, the first notable SNP chip study to canvass a good portion of the genome for links to complex disease was published in the journal *Nature*. The study, known as the Wellcome Trust Case Control Consortium, examined two thousand individuals for seven common diseases, and researchers reported finding many genetic signals that were promising candidates as triggers for each disease.

Although the GWA studies would find thousands of sites on the human genome linked to various diseases, they would be dogged by questions about whether they were worth the considerable cost. Critics argued that many of the genetic differences found were associated with only a small increased risk for a given disease.

Jacob chose not to get involved in the GWA studies, believing he was better off continuing to use his rats to understand disease, working toward an eventual collaboration on genomic medicine with the nearby hospital. His connections were deepening with Children's Hospital, where the pediatrics department had grown into a force of its own.

When Kliegman had arrived in 1993, it was to a community hospital whose faculty were not internationally known and attracted little federally funded research. A disturbing reality was that board members with sick kids would go to Mayo Clinic or Boston Children's Hospital instead of their own institution. Kliegman set out to address those deficiencies.

He kept the genetics mission front and center with the hospital's trustees, even discussing it while dog walking with Edward Zore, a neighbor and trustee who headed Milwaukee's Northwestern Mutual Life Insurance Company. Consequently, when Kliegman brought on a new department head, the trustees were more involved, more inquisitive. They now asked how much funding the new hire had, showing a new understanding of how reputations are made in the research world. When Kliegman told the trustees about his department's climb in the NIH rankings for research funding, they applauded.

"We changed the culture," Kliegman said.

By 2007, the proof was visible throughout the hospital. Broeckel led Children's first genomic partnership, a strategic alliance with Affymetrix, the maker of the SNP chip that was used to detect single-letter variations. The agreement provided lower pricing on Affymetrix's current generation of chips. Children's Hospital researchers would use the chips to study the DNA of a large number of the hospital's patients.

A new $140 million Children's Research Institute building, which housed laboratories and more than eight thousand rats, had just opened. Francis Collins, the director of the National Human Genome Research Institute, gave the inaugural speech. The hospital's pediatrics department had ranked forty-fourth in 2003 in National Institutes of Health research funding. Now it ranked twenty-third and was poised to break into the top twenty in 2008.

At the same time, genomics was changing. No longer the province of a few pioneers in a small corner of the scientific world, as it had been when Jacob took that walk along the beach with Cowley, the field was now burgeoning.

In May 2007, Craig Venter, the scientist who had set up a private effort to compete with the Human Genome Project, completed the sequence of his entire genome. Later that year, Venter was just the third American and one of only a handful of scientists ever chosen to deliver the prestigious Richard Dimbleby Lecture, sponsored by the British Broadcasting Corporation. During his talk, Venter declared that the future of society depended upon its ability to understand and use DNA.

Days after Venter transmitted his genetic script to the national database, in a ceremony at Baylor College of Medicine in Houston, Jonathan Rothberg, the founder of 454 Life Sciences, presented James Watson with a portable hard drive containing the sequence of Watson's genome. 454's next-generation sequencing machine had done the job in thirteen weeks for $1 million.

That was still prohibitive for everyday use but much reduced from the seven years and $600 million it had taken to sequence the first human genome less than a decade earlier. With computer processing speeds continuing to accelerate, the cost of sequencing was plunging. The $1,000 genome was no longer a wishful goal; it had become an inevitability.

Although Jacob does not have the resources of the big sequencing centers such as Baylor, Eric Lander's Broad Institute, and Washington University in St. Louis, he knows how to keep his program competitive. He is preparing to buy a next-generation sequencing machine that will do the work of roughly two hundred of the machines that had been the Human Genome Project's workhorses. By mid-2008, his lab will have a 454 machine—the same model that sequenced James Watson's genome.

Chapter 9

Patient X

FEBRUARY 2008–AUGUST 2008

Half a dozen doctors now form the core of Nic Volker's medical team. After work, they take home their questions about his case, playing out hunches, searching the medical literature, calling and emailing doctors at other hospitals late into the night. At conferences, they mention Patient X, the boy with the mysterious disease. They ask colleagues if they have seen anything like this, if they have any ideas at all what the disease might be. The mystery only intensifies.

"There were no other patients described with his physical condition," says Bill Grossman, the immunologist.

This is the flip side of the diagnostic odyssey consuming Nic's family: the doctors are on their own odyssey.

Whenever the journey stalls, they return once again to the Crohn's disease theory. And again they find themselves defeated by the disease's refusal to respond like Crohn's, to improve with treatment as Crohn's does.

A more general theory among the doctors holds that Nic has some kind of autoimmune disease in which the immune system turns against the body, killing healthy cells. Grossman spends countless hours dissecting complex genetic pathways, groups of genes that influence one another by suppressing some vital processes and triggering others. They are in some ways like lines of dominoes toppling into one another or acting to prevent such a cascade.

Grossman hopes these genetic pathways will provide a window into what is going on inside Nic. The doctor orders tests of various individual genes and tests of Nic's immune system. Medicine has come to depend on such tests. They offer tiny snapshots of what is happening inside a patient's body rather than casting a broad net over the whole genetic script. They work, if a doctor knows where to look and knows which gene is likely the source of a problem. But if the doctor is uncertain, as Grossman and all the other doctors on Nic's case are, the tests are no more than educated guesses.

"He went through the whole gamut of everything we had and then some," Grossman would say. The immunologist with a passion for the hardest cases would be forced to admit, "The more we got back, the more confusing the picture was. . . . He was really a one-in-a-billion kind of case."

The theory that Nic's immune system is the problem does raise the possibility of a treatment. If there is something fundamentally wrong, causing his cells to attack his body and bore holes into his intestines, the answer might be to give Nic a new immune system delivered through a bone marrow transplant. In this scenario, doctors would wipe out Nic's faulty immune system by bombarding him with powerful chemotherapy drugs, the kinds used to destroy cancer cells. Then they would inject a donor's cells into Nic's bone marrow. Bone marrow is the spongy tissue inside bone that produces blood cells. The transplanted cells would help Nic's body build a brand-new immune system, one without the defect. That is the hope, at least.

Such transplants are fairly common but still risky. In the transition period between the old and new immune systems, Nic would be vulnerable to any infection or virus. His body could reject the transplanted cells. The treatment, in Grossman's view, carries a 50 percent chance of working; it also carries a good chance of doing him harm, even killing him.

The Medical College and Children's Hospital have experts in transplant medicine, but when they discuss the bone marrow procedure for Nic, they cannot approve it. The procedure is not to be taken lightly, and they cannot recommend it for a child who has no diagnosis. It makes no sense to forge ahead with a dangerous treatment when they do not understand the problem they are trying to fix.

"If you transplant certain defects, you can kill a kid," Grossman explains. "You have to know what you're treating."

Nic is trapped in another medical catch-22: Children's Hospital has performed every test on him and tried just about every treatment, and the bone marrow transplant is the one thing left. But doctors cannot attempt it unless they understand the illness. In October, when Nic goes home, they have merely bought time.

Three months later, the cycle of desperation begins again. In February 2008, Nic returns to Children's Hospital. His rectum hurts so much he screams. A different section of his colon is now diseased. Doctors take Nic's blood and tell Amylynne that the disease is no longer in remission; it has reawakened.

A year and a half into Nic's illness, Amylynne and Sean begin to understand the full weight of their predicament.

> *February 11 . . . the immunologist believes again that Nicky has something that has never been seen before, no reported cases in the U.S. or world. . . . Nic's prognosis has changed back again to "unknown." The doctors are at a loss again of what to do with Nicky. They have never seen a case like his and some have said they are just guessing on what to do.*

A month later, the mysterious disease has spread. The doctors at Children's recommend that the Volkers seek what will now be a third opinion from a specialist in Cincinnati. But more bad news follows.

In May, Nic's weight drops again. There are spells when he will not move from the family couch. The hospital in Cincinnati cannot take him right away; the earliest they can see him is late summer. Moreover, the Volkers are having problems receiving the necessary insurance approval for the trip to Cincinnati.

In June, Nic is admitted once more to Children's Hospital of Wisconsin. He is very weak. He needs a blood transfusion, but the medical staff fear he might go into cardiac arrest during the procedure. He is now more than three and a half years old. He weighs just 19 pounds; the average for his age is 34.

Doctors tell Amylynne her son now has something called "refeeding syndrome." The syndrome was first described after World War II when Americans who had been starving in prisoner-of-war camps in Japan were reintroduced to food. The sudden addition of food after severe malnutrition led to metabolic problems. The ex-prisoners even developed brain, heart, and lung complications.

The diagnosis confirms what the Volkers have seen in the latest grim evolution of Nic's disease: he has the distended belly and bony limbs of a famine victim. One month at the hospital stretches into two. His temperature climbs above 103 degrees. He has trouble sleeping. The doctors test Nic's bone marrow; but, like the other tests, this only tells them what the boy does not have. No virus, no sign of cancer. The doctors can offer only guesses.

During that year, 2008, longing to have a broader impact on children's health, Grossman leaves the hospital in Wisconsin for the pharmaceutical company Merck. It has taken all of his skill to treat Nic's disease, and it was not enough. The disease is winning.

"It was hard leaving all of my patients," Grossman recalls. "Nic's case was one of the hardest ones. He was one of the kids in the worst shape when I left."

In August, Cincinnati Children's Hospital Medical Center is finally ready for Nic. The hospital is known across the country for
its work on inflammatory bowel disease. Lee A. (Ted) Denson, the
medical director of the Inflammatory Bowel Disease Center, has
spent years studying what drives the illness at the molecular level.
He has an expertise in growth failure, one of the first symptoms he
notices upon examining Nic.

Denson and his colleagues conduct a battery of blood and genetic tests. But Nic's illness stumps them just as it has Grossman
and the doctors in Milwaukee and Madison. They, too, can find no
specific genetic cause.

"He's certainly one of the sickest children we've seen," Denson
will recall a few years later. "You needed to give him about twice
as much nutrition as you'd normally give a child to get any weight
gain at all."

As best the Cincinnati doctors can tell, Nic has an immune disorder. For some reason, white blood cells called lymphocytes have
kicked into overdrive, destroying the cells that line the intestinal
wall and causing severe inflammation. The Volkers return to Wisconsin with a recommendation from the hospital in Cincinnati:
Nic should receive more nutrition until he is well enough to have
his diseased colon removed. Removing the colon would eliminate
the center of all that inflammation, the source of so much of Nic's
pain.

The idea of removing the colon, however, fills Amylynne and
Sean with dread; it would be a permanent step. It is hard for the
Volkers to imagine their son losing a critical part of his digestive
system. But the fearful permanence of taking out Nic's colon is
more of a psychological barrier than a medical one.

The truth is that his bottom and intestinal system have been
so badly damaged by the disease that it is unlikely Nic will ever
regain the functioning he had before the disease struck. The bag
beneath his shirt, where his waste accumulates, the bag he hates,

is no temporary discomfort; it will be with him for the rest of his life.

The question now is whether doctors can figure out Nic's disease before it exacts any further price; whether Nic can heal and grow and return home the healthy, wildly energetic little boy he once was.

Chapter 10

No More Secrets

2008

When Nic was still at the hospital in Madison, Amylynne had been content to give doctors wide latitude in the decisions they made about her son. After all, they were the experts and she was just a mom who lacked the stomach for dealing with blood and other bodily fluids. But as Nic's illness wears on, Amylynne finds reason to grow bolder with his medical team.

The way she sees it, Nic's doctors and nurses too often leave her in the dark. She wants to know—needs to know—more about his progress and treatment. Despite the doctors' efforts, Amylynne feels she is losing her son. The more treatments she watches fail, the more convinced she becomes that she knows as much as the doctors do about Nic's disease. Granted, they understand the tests better, and they have seen far more examples of how diseases progress. But they are stumped. None of the doctors, in her view, sees as much of Nic's disease as she does. She is the one who can never take a break from it. She is certain the doctors are making a mistake by not involving her more.

And there is another issue: Amylynne suspects that the doctors and nurses do not take her seriously. Many of her routines in the hospital, the makeup and business suits, the endless hours gathering Internet research, are meant to convince the medical staff that she is serious, smart, worthy of being included in discussions about his care. Yet Amylynne feels she is making no progress. She still feels left on the sidelines, someone to be seen, watched, managed, but not listened to.

One day a nurse, not one of her favorites, pulls Amylynne aside and hands her a piece of paper. Written on it is the name of a breast reduction surgeon. Amylynne might want to pay him a visit, the nurse suggests. The stunt angers her, but just for a moment; then it gets her thinking.

Amylynne has long had a hunch that the nurses say things behind her back about her large breasts. She believes she makes some of the doctors uncomfortable. She is convinced that they judge her. Whether or not it was intended as a cruel barb, Amylynne takes the nurse's suggestion seriously. It's frustrating to realize that after all the care she's taken, her appearance could still play such a role in her son's care. But breast reduction surgery might actually get them all to take her more seriously, to include her in conversations about Nic's treatment, instead of shutting her out.

Amylynne's father had, in fact, proposed such a surgery when she was still a teenager. She'd rejected it then, but now it has appeal. The logic for getting the surgery in the midst of her son's own medical drama might seem curious, but to the mother, it is about doing anything that could possibly improve Nic's chances of survival.

It is easy to feel powerless in a hospital. If the surgery will make her feel less helpless, less disenfranchised, she has no difficulty taking that step.

Two years into Nic's mysterious illness, Amylynne has learned she cannot control much of what is happening to her son: his inability to eat food, the intestinal holes, the long stretches in a hospital room away from home.

She can, however, be an advocate for Nic. She can do her own research, gathering information from the web and opinions from the doctors in her family. And if Nic's doctors and nurses don't view her as a member of the team, then she will change the image that they see. They already see a woman who manages the look she projects—what she eats, how she works out, the clothes she wears in the hospital. Amylynne decides to make breast reduction part of the picture.

Taking such a decisive step is partly a response to the helplessness she felt during Nic's sepsis crisis in 2007. So much of that night was a blur for Amylynne. She didn't realize at the time how panicked the staff was. She didn't understand the ominous signals from the monitors. She knew her son was very sick, but it didn't occur to her that Nic might actually die.

What kept her calm was the care of a very skilled nurse named Faith, who was working with her son. She thought the nurse's name might be a sign that Nic would pull through. But then, early the next morning, Nic was suddenly transferred to another ward and Faith was gone. The staff was brusque. The move had been decided. Amylynne did not know why. She could only watch it happen.

The change was abrupt, not just for her but for Nic. She viewed the transfer as a factor in Nic's health, a small thing, maybe, but one that might confuse and demoralize him, weakening him as surely as the wrong medication.

It is clear to Amylynne that she and her son are on someone else's turf. With tens of thousands of patients admitted each year, the hospital has rules and procedures in place to make sure everything runs smoothly. But Amylynne has come to believe that many of those rules and procedures are there to make things easier for the doctors and nurses, not for the patients and their families.

When you live in a hospital, you are an open book. The forms you sign and the medical records you generate tell everyone about your child, your family, your history. Doctors and nurses and other staff are in and out of your son's room all the time. There is no privacy. You step away to go to the bathroom, and when you emerge someone new is standing there, looking Nic over, asking questions, taking blood without an explanation. It only seems right that if the medical staff has a free pass into your life, you should have the same access into their decision making.

But Amylynne has no such access. Sometimes she learns about the doctors' latest theory from the nursing staff. Sometimes, it seems, she is the last to know. There are moments when she is taking a short break from the hospital, resting across the street in her room at the Ronald McDonald House, a home away from home for families of the most seriously ill children. Then a doctor decides to change a medication, or Nic's condition takes a sudden turn for the worse. All at once there is a strange voice on the phone telling Amylynne what has already happened. The mother finds herself constantly struggling to process the rush of new information.

The course of her son's care sometimes seems like a secret, and Amylynne has an aversion to secrets that goes back many years.

As a child, Amylynne was forced to keep a big secret. Her mother, Nedra, left the family, first for a month, then for much longer periods through Amylynne's childhood. By the time Amylynne was in kindergarten, Nedra had her own apartment. Her mother and father continued to date off and on but finally divorced when Amylynne was twelve.

Yet even after the divorce, it seemed to Amylynne that the couple did everything they could to avoid allowing the messy truth to poke through the tidy image of an intact family. Although

Amylynne's mother lived elsewhere, she would come to the split-level, white-columned family home in Monroe to entertain guests. Nedra was there when Doc Severinsen played at a party in the backyard. She joined her husband and children for dinner at the local country club. She would sometimes take Amylynne to art classes in Madison or to her rehearsals when she had a part in another avant-garde theater production. Nedra encouraged her children to call her by her first name rather than "Mom." She wore caftans and gaucho pants, thousand-dollar outfits from Bloomingdale's and Neiman Marcus, which she acquired during shopping trips to New York and Chicago on her friend's private plane. She was liberal and agnostic and thoroughly unlike any of the other mothers in Monroe. With her Mohawk-like hairstyle and long, polished nails, Nedra inspired Amylynne's friends to write a poem about the "Dragon Lady."

Though Nedra came and went, the divorce was kept secret, even from members of the extended family. When the Santiagos traveled back to the Philippines for a relative's fiftieth wedding anniversary, Amylynne recalls that she and her three older siblings were instructed not to mention that Mom and Dad no longer lived together.

With her friends, Amylynne never discussed the divorce or the unusual arrangements at home. Those who visited noticed only that her mother was not around much.

It fell mostly to Fernando, Amylynne's father, to fill the role of both parents. Amylynne attended Mass at the hospital chapel with him, then accompanied him on his rounds. When she skidded off her skateboard, he sewed up the wound at home. When she lost her house key after school, she would wait for her dad at the clinic, watching him burn off warts, stitch cuts, and perform other procedures.

A Spaniard with dark, wavy hair and impeccable style who was raised in the Philippines, Fernando gave Amylynne her fashion sense. He was embarrassed when his medical colleagues came up with an odd nickname for his teenage daughter: Dolly Parton. The nickname came up during meetings, and it bothered him. It was one of the reasons he encouraged Amylynne to get breast reduction surgery.

There was no question that Amylynne's breasts made her stand out. She had used that to her advantage during her twenties while living in Hawaii. She entered bikini contests, modeled swimwear for Sears, and earned as much as $4,000 a weekend as Miss Smirnoff, entertaining wealthy business people on the company yacht. Professional football players asked to have their pictures taken with her. She leveraged her modeling into a marketing business that worked with Pro Bowl and beach volleyball events.

But all these years later, she concludes that a full figure offers no such advantages in the hospital. When Amylynne enters an exam room with Nic, walks with him down a corridor, or visits him in his room, she senses where everyone's eyes go first.

She has no back pain, no reason for the breast reduction except the hope that it will help her become a more effective advocate for her son. If it improves her relationships with doctors and Nic benefits, she will worry about any other consequences later.

She decides to do it, and in the spring of 2008, after going through the approval process, Amylynne drives herself to a Milwaukee hospital to undergo the surgery. It is an outpatient procedure. At home the next day, as the anesthesia wears off, Amylynne hunches over in pain. Moving around and lifting Nic only increase her discomfort.

Although her daughters are aware of the surgery, Amylynne gets no reaction from them. Sean is not entirely happy with the result, but Amylynne is. The change at the hospital is immediate and gratifying.

"Everybody treated me differently," she recalls.

Amylynne starts to treat the doctors and nurses differently, too. The more she has thought about it, the more convinced she has become that she needs to take matters into her own hands. She ought to be consulted before any important change is made. This is not something she intends to ask; it is something she will demand.

Now, in 2008, when doctors suggest a new medicine or a diagnosis that has not been discussed before, Amylynne is more aggressive, questioning their reasoning with a new confidence. When she is not happy with Nic's treatment, she says so. She calls meetings when she deems it necessary, contacts hospital executives, even hires and fires doctors.

Why not? Some doctors and nurses are better than others. Amylynne intends to get the best for her son—whether the medical staff like her approach or not.

She believes she must be on her game all the time. Once, thinking Nic might benefit from hyperbaric treatments, Amylynne decides to contact Harry Whelan, a pediatric neurologist and director of Children's Hospital's hyperbaric unit. Hyperbaric treatments take place in a room in which the air pressure is three times higher than normal, stimulating growth factors and stem cells inside the body that can improve wound healing and provide other health benefits. Unable to schedule an appointment with the busy doctor through normal channels, Amylynne finds another way. Driving near the hospital one day, she spots a sweaty jogger with big, bushy eyebrows running down Watertown Plank Road. She has seen his picture online and is certain it is Whelan. Nic's primary doctor won't like it, but Amylynne pulls over anyway, rolls down the car window, and gets her appointment.

On another occasion, when Amylynne doesn't like the way a nurse is trying to insert a tube up Nic's nose, she politely but firmly tells her what she is doing wrong and asks her to leave the room. Leilani, who is visiting her brother at the time, thinks the nurse seems terrified. Later, Amylynne reminds her daughter that she was polite, saying "thank you" before ordering the nurse to leave.

Despite all of the mother's efforts, there is little change in Nic's outlook. He passes much of 2008 in the hospital. Amylynne becomes increasingly disappointed with his primary doctor, Michael Stephens. She does not feel as if she hears from him or sees him enough. She often feels as if Stephens has cut her out of the loop, leaving her without a clear sense of what is happening and which treatment plan is being pursued. She considers bringing in another doctor.

One possibility is Alan Mayer, a short, intense pediatric gastroenterologist with dark hair and a kind but firm manner. Nic and Amylynne have met Mayer before when he was the doctor on call.

As she does with all the doctors, Amylynne researches Mayer on the Internet and assembles a rough profile of his qualifications. She is impressed that he has a PhD as well as an MD. He treats patients but also does research; the research experience might be useful in a case as complex as Nic's. Mayer received his degrees from Cornell University and earned fellowships at Children's Hospital of Philadelphia, Massachusetts General Hospital, Boston Children's Hospital, and Harvard University. Amylynne learns his background so well that years later she can recite the schools and hospitals without pause.

Despite his impressive resume, Mayer is soft-spoken. He tells Amylynne what he is thinking. He listens. And he is good with Nic. Unlike some of the other doctors, he tries to establish a rapport.

But the first two times she approaches him to take over her son's care, Mayer makes no commitment.

Nic is now a few months past his fourth birthday. He still weighs less than 20 pounds. In 2008 alone, he made twenty-eight visits to the operating room—and he is headed back yet again.

Amylynne finds herself stuck in a place she'd never expected to be, this cycle of uncertainty, of temporary treatments without a long-term answer. She had never imagined the procession of doctors and nurses, never considered that none of them would be able to tell her what is wrong with her son. Nic's disease is a secret of the worst sort.

But Amylynne has changed. Now Nic's care needs to change, too. She prays for something to break the cycle, for someone new to enter the picture.

Chapter 11

Survivors

FEBRUARY–MARCH 2009

Alan Mayer goes over the request again and again, wondering what he should do. Nic Volker's mother has asked him to take over the care of her son. The boy and his mysterious disease might be the challenge of a career. But the other doctors Mayer consults are not encouraging. Their message boils down to this: The child is a lost cause. He's going to die, and you will be blamed.

No one wants to be burdened with this child's death—that's how it seems to Mayer.

The doctor feels deep sympathy for Nic, a boy who has spent so much of his short life in the hospital, who has fallen victim to a cruel disease, who is rapidly exhausting all that modern medicine has to offer, becoming a humbling example of its limits.

Mayer's sympathy for Nic is balanced by a wariness of Amy-lynne.

Through the hospital grapevine he has heard that this mother does not respect boundaries between parents and medical staff, that she has manipulated, antagonized, pushed, argued, even reduced some nurses to tears.

Later Mayer will put it this way: "I was duly warned."

But he can be formidable, too. Like Amylynne, Mayer is tough and stubborn. Like her, he is a student of people.

The son of two Holocaust survivors, Mayer has a fascination with fate and probability. His father survived for a year and a half hiding from the Nazis in a bunker beneath the basement of a house in Drohobych, Poland, now part of Ukraine. The only ventilation in the bunker came from the house's fireplace and the town sewer system. Before the war, Drohobych had been home to about 17,000 Jews. By the war's end, only 150 had survived, 50 of them in that bunker.

Mayer's mother lived in Vishkov, near Warsaw. With some members of her family, she escaped the Nazi invasion by fleeing into the part of Poland under Russian control. Her father was killed by the Germans within days of the invasion. But she and her surviving relatives were transported to Siberia, where they remained for the duration of the war.

During his childhood in suburban New Jersey, then south Florida, Mayer had grown up with the stories of how his parents lived through the Holocaust. Because he showed a gift with numbers from an early age, excelling in math and science, he sometimes considered the long odds associated with his parents' survival and his own existence. He was well aware of his incredible good fortune simply to be alive—and with that came a burden. His whole life, Mayer would feel he had a responsibility to make the most of the chance he'd been given.

"Early on you feel like you're on this earth for a purpose," he says. "You feel extremely fortunate because of the infinitesimal odds that your parents would survive all of this. Most of your friends, their parents met at college. They had no threats to their existence. But extremely large and powerful forces were aligned against you."

As a child, he liked to take things apart. Mayer's younger brother, his only sibling, received a remote-control police car one year as a birthday present. After three days, Mayer finally found an unguarded moment to spirit the car away, pull it apart, and figure out what made it run. As he grew older, Mayer gravitated toward science and medicine. He would strive to keep a foot in both worlds.

Much as he loved science, the field had its frustrations. One discovery opened the frontiers to more discoveries, and it was easy to feel as if anything you discovered would have been found sooner or later by someone else. Research could, and often did, crawl along. Medicine offered a compelling balance, that missing sense of urgency. Victories in a hospital were more immediate, more personal, and easier to define; patients needed help now, not in twenty years.

"Medicine," Mayer says, "has its moments when you can go home at the end of the day without any ambiguity, knowing you helped somebody."

The losses, though, were heartbreaking, humbling reminders that doctors are neither supermen nor gods. It was a lesson Mayer came to appreciate during his years of medical training at Children's Hospital of Philadelphia and Boston Children's Hospital. The better the hospital, the more likely it is to attract patients with the most horrific and mystifying illnesses.

"You develop a lot of humility when it comes to disease," Mayer says. "At smaller places you can develop hubris. At Boston, kids come from all over the place with diseases that could not be treated elsewhere. You become small in the face of disease."

When children died, it was more than humbling, especially as Mayer and his wife became parents themselves, raising a boy and a girl. Lacking the tools to halt an illness and save a child became more than an intellectual frustration. It made him despair. When a child perished, he put himself in the position of the parents. He tried on their sadness, wearing it like a lead suit. Feeling each death this way made him realize he could not work in pediatric oncology, where every child on the floor is facing death.

Instead, he specialized in gastrointestinal medicine, a good fit for a doctor who felt humbled by disease. The gastrointestinal system is complex and not well understood.

Mayer's tinkering side also suited his work. That boyhood fondness for taking things apart and studying the relationships among the parts translated into a fascination with genetics. When he looked at how diseases unfold in the body, he found himself examining the basic molecular instructions—instructions that come from genes.

Mayer's two chief interests—genetics and the gut—led him to fellowships at Massachusetts General Hospital and Boston Children's Hospital. At Mass General, he worked in the lab of Mark Fishman, a pioneer in the study of zebrafish who would go on to become the president of Novartis Pharmaceuticals' Institutes for BioMedical Research. Zebrafish are a popular model organism, especially for work examining the function of genes. Mayer focused on the zebrafish gut, homing in on the genes most important to the digestive system's development.

A few doors down from him at Mass General was an office that had, until very recently, belonged to a rising young genetics researcher who'd also worked for Fishman, although not in his lab. Howard Jacob, who'd been involved in creating the genetic maps of both rats and zebrafish, had been at the hospital when Mayer interviewed but had left for Wisconsin just before the new doctor arrived. Mayer knew Jacob by reputation and found himself following Jacob's footsteps in some respects, even using some of the same lab equipment Jacob had used in his research.

After a few years of postdoctoral work, Mayer, too, began looking for better opportunities. He had been teaching at Harvard Medical School while continuing his research. Though he had hoped to stay at Harvard, he was now unable to advance to the point of receiving his own lab there. It was 2003, and Mayer and his wife had a young son and daughter to consider. If advancement meant leaving Harvard, Mayer planned to aim for a position at Vanderbilt or at Washington University in St. Louis, one of the nation's foremost centers for the study of genomics.

At the same time, however, he found himself recruited by a dark horse, the Medical College of Wisconsin. Colin Rudolph, the director of the Division of Pediatric Gastroenterology and Nutrition at the college, wanted a researcher for his department. A colleague of Mayer's had recommended him, but Wisconsin was not exactly the kind of place the young doctor had been considering.

"I'd been at U Penn and Harvard," he would recall. "The likelihood I was going to entrust my career to a small medical college in Milwaukee was remote, to say the least."

Vanderbilt and Washington Universities offered him standard packages. The surprise turned out to be the Medical College of Wisconsin.

"They were almost irrational in how generous they were to help me build my lab," Mayer would recall. "They were willing to give me much, much more than anybody else."

The Medical College committed more than $1 million to his lab. When he began looking into the area, Mayer was impressed by the suburbs north of Milwaukee, which offered good schools and affordable housing.

Colleagues in Massachusetts who'd heard that he was considering a move to Milwaukee told Mayer he was making a big mistake. A smart young doctor with a bright future needed a strong environment. Otherwise, it would not matter how smart he was. It was the same message Howard Jacob had heard before he'd left Boston a few years earlier. Like Jacob, Mayer chose to ignore it, though for different reasons: he was attracted by the prospect of an institution that could support not only his clinical work but his research.

At the Medical College, Mayer would be able to nourish both sides of him: the doctor and the scientist. Like many at the Medical College, his job was a joint appointment with Children's Hospital of Wisconsin. When he started in 2003, research consumed the vast majority of his time, about 90 percent, with the other 10 percent going to his clinical work at Children's Hospital. He continued his study of zebrafish and the genes involved in gut development. It was just the opportunity he'd hoped it would be, at least for a while.

Between 2007 and 2008, however, the research climate began to change. Government grants began drying up. The odds of receiving funding were on their way down, from 30 percent to around 10 percent. At the same time, as people lost jobs during the economic collapse, hospitals began to receive more reimbursements from Medicaid, rather than from private insurance companies. The smaller Medicaid reimbursements left hospitals grasping for ways to bring in more revenue.

In February 2009, the Medical College informs Mayer that money previously committed to his research will not be available. A few existing grants are unaffected by the change, allowing him to continue the lab work for another year. But Mayer, who started doing research at seventeen and had his first paper published at twenty, now realizes that his work in Milwaukee is destined to be almost entirely clinical. The research he values so much will be over.

In response, he begins searching for other jobs in other cities. Then, in the midst of his job search, he is presented with the clinical challenge of his career: Nic Volker.

In at least one respect, Mayer has been well prepared for the role. One of the most crucial lessons he has learned in science, and especially in working with abnormal zebrafish, is to recognize when things don't fit. People become wedded to paradigms. Sometimes, however, the paradigm is wrong. Nic Volker is the ultimate case; he fits none of the prevailing theories.

As a doctor, Mayer likes caring for children, especially those with challenging ailments. But he has never had a patient like Nic. Or a parent like Amylynne.

For two weeks, Mayer considers whether or not to take the lead in Nic's care. The turning point in his deliberations comes on a dreary March afternoon in 2009, when he attends a care conference with the mother and several of Nic's doctors. Mayer keeps silent for the first forty-five minutes, listening as the other doctors discuss Nic's situation. He still has numerous holes in his intestines. They keep leaking stool onto the wound that will not heal. The doctors suggest approaches that have been tried already. Perhaps they need more time to work. There is more talk of a bone marrow transplant, but that runs into the same obstacle as always: the problem with Nic's immune system does not correspond to any known defect, so there is still no diagnosis. Without a diagnosis, the transplant doctors cannot justify approving the risky procedure. Even if they could overlook the lack of a diagnosis, there is one factor doctors cannot ignore: Nic is not healthy enough to survive a transplant.

After listening to the others, Mayer offers his view. He has just examined Nic's gastrointestinal tract, and he says it is clear that they have failed to get the inflammation under control. They will need to try something new, perhaps different medications. If they are still not successful, they will have to consider surgical options— the main one at this point being the removal of the child's colon, the permanent step that still troubles Amylynne. Finally, Mayer says that the transplant doctors must decide precisely what they will need in order to approve a bone marrow transplant.

His overriding message is that the status quo is unacceptable. They need to do something different. Mayer knows that Nic's colon will probably have to be removed. But he watches Amylynne, and senses correctly that she is not ready to accept this step.

How well Mayer has read Amylynne can be seen from her account of the conference in her CaringBridge journal. It is clear that much of what she hears from the other doctors leaves her feeling not hope but dread.

"The conference was hard," she writes. "Some Docs seem so callous and threw out the D word at least 12 times—and the term D's Door at least two times. I have always disliked the D's Door analogy. . . . Please pray against the negativity and the D word."

It is her impression that the doctors "just want to speed ahead" with removal of the colon, known in medical parlance as a colectomy. Here Amylynne breaks her prohibition on the D word. She and Sean, she writes, are "deathly afraid" of the colectomy. The tone of the doctors has given the Volkers the distinct impression that most do not expect Nic to survive the procedure. The mother's instinct is to apply the brakes, to make sure there is no other option before taking this irreversible step.

There is one doctor at the conference—and only one—whom Amylynne praises in her journal.

"He had the most original and thoughtful ideas today," she writes, "and stepped in on my behalf (before I even spoke up one word—I know most of you can't believe that) and told all the other Docs that Nic is my child and it will be my decision what therapies he will receive. . . . I so needed someone to support me and speak up for me and Nic."

That doctor is Alan Mayer. After the conference she approaches Mayer once again about taking over her son's care. By then the doctor has weighed his decision carefully. As wary as he is of the stories he has heard about the mother, he knows he cannot say no to the child. His research is tailing off; his best efforts will now go into patient care. He can focus on Nic.

Mayer tries to be clear-eyed in his view of the case. There are

only two ways the story of Nic Volker can go. Either he will get better, or he will die. There is no middle ground for this child. Despite his training and skill, Mayer is not certain he can save Nic. All he knows for sure is that Nic's unconventional illness will require an unconventional treatment. The boy's disease is formidable—the kind that makes a doctor feel small.

Mayer begins by speaking with Michael Stephens, Nic's primary doctor. He tells his colleague that he's been asked by Amylynne to take charge of Nic's care: Are you OK with this? he asks. The situation could be awkward, but Stephens has worked in the hospital long enough to understand the mercurial nature of the doctor-parent relationship. Also, Mayer asks if he can continue to consult Stephens throughout Nic's care. Stephens accepts the change with grace, as Mayer will recall. If Stephens harbors any hurt feelings, he keeps them to himself.

When it comes to dealing with Amylynne, Mayer is careful, even calculating. He never discusses boundaries with her. Instead, he decides how their relationship must work. Other doctors have called Amylynne by her first name. He will call her Mrs. Volker. They will be formal with each other. He will talk to her—but on an as-needed basis. He will plan each of their conversations beforehand. *What needs to be accomplished? How do I phrase this?* Mayer never states or explains the ground rules to Mrs. Volker. He simply applies them.

Later, he will explain the relationship this way: "The more difficult the case, the more distance I keep between myself and the parents and family. I might even seem cold. It doesn't serve the patient for me to get emotionally involved with the family."

Mayer believes that Nic is a victim. He is the victim of a horrible disease that no one has been able to treat, but, in a more complex way, he is the victim of a medical system ill equipped to resolve conflicts between staff and families. He knows Amylynne loves her son, yet he believes that her advocacy sometimes goes

beyond what is acceptable to the medical staff. The problem is that a doctor and a mother look at the same medical system from very different vantage points. When Amylynne objects to Nic's transfer from one part of the hospital to another, she sees it as fighting for his safety, preserving care that will give him the greatest chance of survival, and keeping him with the nurses and staff who know him best.

In Mayer's view, Amylynne is only compromising Nic's care by alienating the staff—sometimes to the point where they no longer feel strongly motivated to fight for the boy. "There is a vast gray area in caring for someone in terms of what's required and what can be done," he explains. "When you antagonize people, they're still going to do their jobs, but that's it."

As he sees it, the mother has become like one more organ of Nic's that has to be considered in his care. The heart, the kidneys, the colon, and Amylynne.

In late March, when Mayer takes charge of Nic's care, the boy is still four years old. His fifth birthday is more than six months away. The outlook has again turned grim, even for the mother who refuses to write the D word.

"As far as they can tell at this point, Nic has failed all medical and surgical management," she writes in her journal, "and the only possible alternative is major surgery to remove his colon. . . . The colectomy could be fabulous, don't get me wrong—could make the disease go away—and give Nic's body the ability to absorb food again, normally through the gut. But the downside is he has to heal from this major surgery—and if he isn't healing now—wound sites aren't closing—there is no guarantee Nic will be able to survive the surgery. Or survive past the surgery."

Chapter 12

The Dragon

FEBRUARY–JUNE 2009

Alan Mayer has never faced a challenge like Nic's disease. He learns quickly that this is one case in which modern medicine plays a passive role, reacting to an unpredictable foe. When Mayer takes over Nic's care, he sees that the disease is advancing through the child's colon, ravaging it with holes. To make matters worse, the surgical wound in his abdomen has worsened and poses an immediate threat of infection. Mayer cannot focus on the search for a long-term solution. Instead, he must keep sending Nic back to the operating room to have his wound cleaned, a task that falls to surgeon Marjorie Arca.

Cleaning and dressing a wound is not normally considered surgery. But in Nic's case, each cleaning takes at least two hours. There is no question of allowing him to remain conscious; the pain would be excruciating. Arca would not have a child go through that. So Nic undergoes anesthesia, which carries its own risks. There is always the terrible fear that he will not emerge into consciousness, that he will instead drift into a coma. The wound cleanings are full-blown procedures, and Amylynne worries about them. Yet in time the procedures become routine, to the point where it is sometimes hard to muster the appropriate parental fear.

The routine has a way of masking the strangeness of Nic's situation. If the disease is unlike any the doctors have seen, so is the child's wound. No article in a medical journal or textbook prepares a surgeon for such a wound. It has become so serious that Arca has a standing daily appointment with Nic in the operating room. She finds it easier to schedule the appointment and then cancel in the rare case that it turns out to be unnecessary.

Wound-cleaning days follow their own choreography. Mother and surgeon stand on opposite sides of the line at the entrance to the operating room. Amylynne passes her son to Arca. Nic often wears his Batman cape and costume for the occasion, and Arca, knowing Nic loves it, addresses him as Batman.

Arca has a way with children. She sees the world through their eyes. So when she talks with Nic, she treats his illness like a cartoonish troll or evil stepmother.

"Bad colon," she hisses.

"What would you like to listen to?" she asks. "Jonas Brothers?" Nic is fond of the Jonas Brothers. His mother often hears him singing a line from one of their songs, "A little bit longer and I'll be fine."

Nic has become so familiar with the preoperative procedure that he helps. He holds out the intravenous line in his chest that is used for medications, fluids, and blood draws.

"I would like milky Versed," he says, using a term he has cobbled together for the anesthetic propofol. It reminds him of Versed, a sedation drug with which he's familiar.

The oxygen he receives during the procedure comes in different flavors.

"Blueberry," he says. He knows he gets to choose.

Though there are few places in his world where he has any choice or degree of control, oddly, the operating room is one.

"I'll hold the mask," he says, referring to the oxygen mask. Sometimes residents try to hold it, and he pushes their arms away.

Afterward, when he is in the recovery room and his eyes begin to open and the bright hospital lights flood in, Nic speaks. He always asks for Mom.

Day after day Mayer examines Nic, updates Amylynne, and sends them back to the operating room. He knows as well as anyone that the wound cleanings have settled into a routine. Nic and Amylynne are in the midst of what will become more than a hundred trips to the operating room in 2009 alone, an ordeal that goes beyond medicine. Mayer sees what the endless days in hospital are doing to Nic's childhood, and he understands why the child is grumpy sometimes, why he is not always happy to see Mayer, why he does not hesitate to show his displeasure.

The boy knows where the operating room is. He knows the different floors of the hospital. The massive building, confusing to most adults, is home. What friends Nic has are here. His mom arranges hospital play dates. But friendships in the hospital are fleeting. Nic learns that friends always leave. They go home. Or they die.

In addition to the loneliness, there is the child's intense desire for real food, for something that does not enter his body in liquid form through an intravenous line. It has been months since he has eaten a proper meal. He tells nurses about the steak he will eat when he goes home. At night, he sleeps cradling a bag of Bagel Bites.

There are days he asks his mother for prayers; there are other days when he wrestles with faith.

"Mom," he asks, "is God listening? Is He there? Because why would I be here then?"

It is hard for a mother to know how to respond. Sometimes Amylynne tells him that God is always with us. Sometimes Nic's questions leave her speechless, wiping away tears.

His illness, now entering its third year, has taken a toll on the whole family. When his three sisters can, they visit the hospital, more than an hour away from their home. They think of Nic when they are at home and at school. They find themselves missing the crazy things he says. More disconcerting is the way his illness has left them on their own in Monona. Their mother is rarely ever home. Sometimes their father will decide late at night to drive up to the hospital to see Nic. The girls wake in the morning, and he's gone. Leilani grows fanatical about making sure all the doors are locked at night.

Sometimes there's no food in the house and the girls must do the shopping themselves. Mariah, the oldest, becomes a mother of sorts. She makes the meals and wakes the others for school. The girls no longer eat the fresh produce and healthy fare their mother fed them; frozen pizza and macaroni and cheese become staples. At night, when they have nightmares or just want company, Mariah lies down beside her sisters. They talk about what is happening to their family. They wonder if their parents care about them anymore.

"I knew that what they wanted was to do what was best for Nic," Leilani later says. "But I also felt like they had given up on us, on me."

She copes by praying and reading the Bible. She reads and rereads chapters. Kristen also turns to the Bible; she goes into her bedroom after school and reads until dinnertime. Leilani looks forward to weekend trips to the hospital. It is the one time she feels they are a family.

The girls know their father is hurting, though he seldom talks about it. Sean and Amylynne are living separate lives. They see each other in Nic's hospital room or occasionally at church. Most of the time, Amylynne stays with Nic; Sean stays in Monona, each isolated in private worry.

Sean worries that he is not providing for his family. The financial pressures are relentless. When the family's six-year-old Nissan Altima won't start in the parking garage at Children's Hospital, Sean leaves it there for so long it is towed away. Amylynne pushes Sean to go to the tow lot and retrieve the car, which had cost them more than $20,000.

"I don't have the money to go get it," he tells her.

They lose the car.

It is hard holding things inside as Sean does, but that is his nature. One night he comes home, sits down on the living room floor, and cries as his daughters watch.

"I'm sorry," he says, the only two words he can manage. "I'm sorry. I'm sorry."

Sean endures long absences from his son for the first two years of his illness, taking all the overtime he can get to cover the medical expenses that insurance will not. Early in 2009, when construction jobs dry up, he and Amylynne switch roles. Amylynne takes a job doing clerical work for Great Wolf Resorts, which operates water parks, and Sean begins spending long hours at the hospital with Nic, playing on the floor and talking monster trucks and superheroes. When construction picks up again a few months later, Sean returns to work. At job sites, sometimes his cell phone rings and there is that little high-pitched voice again: "Dad, when are you coming?"

The last days of March 2009 are terrible. In three months, Nic has spent just one day outside the hospital. Amylynne has been planning another, but Nic's temperature soars to 104. He throws up violently, blowing out his nasogastric tube and the IV line for his blood transfusions. The stoma, the opening in his stomach through which waste empties into a bag, swells to twice its normal size. His voice is hoarse. His body trembles with chills.

One of Alan Mayer's first acts as Nic's lead doctor is to deliver the news Amylynne has been resisting. At first, Mayer was suggesting alternatives to removal of the child's colon. Now he is firm. Other options are no longer possible. Nic's colon must be removed without delay to save his life. Amylynne did not want this, but she appreciates Mayer's straight talk.

"It feels different," she writes, "when the Doctor confronts me as a mother to tell me straight out that my precious Nic is better off without his colon than dead with it. And so, there it is, the clarity perhaps that I was looking for, but wasn't expecting in this way."

She is convinced now that Mayer is the answer to the prayer she'd voiced weeks ago.

Gently, Marjorie Arca has also been telling Amylynne that Nic's colon has to be removed. Arca even uses the D word: Nic is dying in the hospital. She needs to remove his colon, but first she must be sure he is healthy enough for surgery.

The major barrier has been the surgical wound in Nic's abdomen that refuses to heal. His healing has been so poor that Amylynne reaches into the realm of alternative medicine, where she finds a product called manuka honey. The honey, found in New Zealand and Australia, is a special formulation created from the nectar of the manuka tree and has been found to fight bacteria. They try the honey. Arca even orders special bandages treated with manuka honey. Nic receives the honey for a month or so, and it does seem to help his wound a little. But the improvement is too slow to justify continuing.

Nic has been in good spirits, drawing bats, assembling puzzles, and rolling balls of Play-Doh. Still, he is aware of what's coming. He says he does not want the operation to remove his colon.

In early April, he is still not well enough for the surgery. Mayer must stabilize him. Nic must regain strength.

Easter comes and goes. The boy receives baskets from the Easter Bunny at the hospital, then at the Ronald McDonald House nearby, where Amylynne lives during the long hospital stays. Doctors allow the Volkers to take Nic to their church in Madison. The pastor and congregation pray over him.

Finally, on April 22, he goes into surgery. Arca removes a colon that has become an ugly mess, its pink hue discolored by an enormous yellow ulcer covered in pus. No part of the colon has been left undamaged.

Nic wakes feeling chatty, telling Arca her surgical glasses look funny.

Around two in the morning, a stabbing sensation wracks his gut. He screams and asks his mother to pray for his "tummy pain." The screams go on for hours. The complications go on for days. Nic endures a collapsed lung, fevers, and night tremors. He sees snakes crawling up the hospital walls, hallucinations brought on by the pain narcotics.

In the weeks that follow, Nic gets an infection, one of the new incisions in his abdomen refuses to heal, and his oxygen levels drop so low that he must receive air through a mask. The surgical opening begins to bleed, and his belly grows distended. He longs to eat a real meal, but snuggles instead with packages of baby beef sticks in one arm, candy in another.

Not all of the news is bad. In her journal, Amylynne writes that she and Mayer are on the same page when it comes to Nic's care.

> *I really like him. He asked my opinion and asked what I thought and where I thought Nic was. And after giving him my thoughts, he reassured me I was being reasonable, and we would wait and see what all the newly ordered labs and scans and X-rays would find first before jumping the gun.*

There are good spells, days when Nic has no infection, when the disease seems to lack fuel. He reaches thirty pounds. He finally enjoys real food, a meal of chicken soup, roasted potatoes, carrots, and pork tenderloin. But the food is the fuel, the on switch for his illness.

There is no escaping the bad days. The pain in Nic's abdomen returns. He receives transfusions of granulocytes, a kind of white blood cell. Here, as in other aspects of Nic's illness, there is a cruel irony. The granulocytes appear to help his wound sites heal. But the doctors want him to stabilize before they give him more transfusions. In essence, he is not well enough to receive the medicine that may make him better.

In May, Nic sings and roars around the hospital in his toy Batmobile. He gets a pass to go visit the Milwaukee County Zoo.

In June, he loses weight and vomits, enduring stomach pain and poor wound healing.

The lack of healing in one of the recent surgical incisions is so puzzling that Mayer talks of referring the child to the National Institutes of Health for further tests. Perhaps Nic has a disease of the connective tissue, a layer that forms a sort of support structure protecting organs.

Mayer has come to think of the disease as a dragon. It is fantastically destructive, almost mythical. If he had not seen the illness for himself, he would not believe such a thing could exist. Using all of his medical skill, Mayer can only lull the dragon into a temporary sleep. It's not enough.

"You want this dragon gone," he says.

At times, Mayer does not even feel as if his work involves skill anymore: it seems more like bailing water from a sinking ship. He can keep the ship afloat, but nothing he does addresses the source of the leak.

The pressure is exhausting: as if he were on call twenty-four hours a day for this one child. When a new attending doctor sees Nic and offers a different opinion about his illness or his present condition, Mayer knows he will be called. Amylynne will want to know why. Why does the new doctor have a different idea? What does he know that Mayer does not?

Even when Mayer is on vacation, there are calls from the medical staff at Children's all the time: Here is the situation. What do you want to do?

"It's hard to talk about," he will later explain, his voice growing quiet at the memory. "You imagine yourself in a situation where a child is about to be lost and you don't have the answers. You don't have anything you can do about it."

Not knowing what he is up against is a terrible, helpless feeling. The dragon follows him home each day. It is in the car each morning as he drives to work, wondering if he will learn that Nic has died. It is with him as he lies in bed awake, trying to figure out its weakness. The dragon never leaves him, even as he drifts into unconsciousness. The case churns through his brain as he sleeps. The doctor dreams of being lost in the woods and wakes in the morning with a tightness in his throat that comes from knowing that he is responsible for another life.

He must return to the hospital. Nic's disease is waiting.

Chapter 13

The Genome Joke

JUNE 2009

The dragon is winning.

Mayer has tried various medications to stabilize Nic. He has restricted all of the boy's nutrition to IV feeding, no exceptions. He has suppressed the child's immune system, careful to make sure Nic is not under attack from a virus, always a concern during long hospital stays. He uses special antibiotics to kill off bacteria, hoping this will signal an all-clear to Nic's immune system—hoping that somehow the immune system will call off its guns and stop attacking the child's intestines.

Despite all these efforts, one of the key symptoms remains unchanged: the sedimentation rate, which measures inflammation, barely budges. Inflammation is the engine of this disease. Still, the doctor observes something interesting. The part of Nic's intestine that has been bypassed in one of the earlier operations has healed. Once food stopped going through that portion, the inflammation stopped, too. He takes this fact as proof of what he's suspected. Deprive the dragon of food, and it lies still. By giving Nic antibiotics and restricting him to intravenous nutrition, the doctor is removing fuel from the disease.

The problem is that the child is too far gone to survive using this strategy alone. He has too many fistulas and his disease is on the move, climbing what is left of his functioning intestine. Now he is developing small intestine disease, an ominous sign.

By removing Nic's colon, Mayer believes, the medical team saved his life. But without a colon to funnel waste from his body, the boy must rely on his ileum, the final section of the small intestine. Eventually, whatever is wrong with his immune system will begin hammering away at the ileum. Nic can live without a colon as long as he has the ileum. If he loses his ileum, he will not survive to adulthood. There will be no way for his food to be absorbed. He will have to depend on intravenous nutrition for the rest of his life, and eventually, as a result, his liver will sicken and fail. It might take as long as five years, but the boy will die—probably before his tenth birthday.

James Verbsky, now the main immunologist on Nic's case, had believed that the boy was suffering from a defect in the gene Il-10. But when he tests his theory, he finds that Nic's Il-10 gene is functioning as it should. This theory, like all of the others, has not panned out, and Verbsky is now out of ideas. Quietly, to himself, Verbsky admits defeat. He accepts that he will not figure out the disease and save the child. He gives up.

Amylynne continues to press doctors about a bone marrow transplant, but the transplant surgeons remain firm. When a bone marrow transplant is not based on a firm diagnosis, the odds of survival are about 50 percent. In good conscience, the transplant doctors cannot approve the procedure—cannot, in essence, pin the boy's life to the outcome of a coin toss. All around Amylynne, the medical staff are saying they have run out of ideas and options.

The mother's frustration peaks.

"So here we are back to square one, with no diagnosis and no clue as to what is making Nic sick," she writes in her online journal. "The days are getting longer and I feel Nic is just languishing in the hospital. Nic has meds in the morning, and again at night, and nothing routine in between . . . they won't let Nic have consistent daily trips out of the hospital, even to the Ronald McDonald House. Nic is getting depressed and agitated in this wait-and-see mode.

"Please pray for his spirits—and maybe mine, too."

Amylynne tries to lift their spirits by convincing a sympathetic hospital worker to bring a small plastic swimming pool from another area into Nic's room. Although the pool is against regulations, mother and son spend hours kneeling beside it, floating toys and splashing their hands in the water. When Amylynne leaves for a day trip back to Monona, the pool disappears. No one will say who took it.

Near the end of June, Mayer takes a look at Nic's ileum. He can see the disease taking root. Now the doctor cannot sleep at night. He knows this child is going to die. And despite all of his knowledge and skill, he will be unable to do more than watch.

In desperation, he goes to see David Margolis, one of the transplant doctors. Mayer is now reduced to begging Margolis to approve a bone marrow transplant. The response is the same: no diagnosis, no transplant. It would be irresponsible to go ahead without comprehending the nature of the problem.

Mayer understands his colleague's position. He is asking Margolis to assume a huge risk. If the transplant fails, Margolis becomes the doctor who killed Nic.

At the same time, Mayer is thoroughly frustrated. There are no other options on the table. We've done every test that can be done, Mayer explains to Margolis. What will it take? What more do you want us to do?

Neither doctor will recall who said it. The idea is proposed as more of a joke than a possibility.

"What are we supposed to do now, sequence his genome?"

At this point, late June 2009, there are no published reports of patients having had all of their genes sequenced in order to reach a diagnosis. There is no estimate for what it might cost. There are no guidelines for how to sequence a live patient. There are many ethical questions that accompany sequencing, but they've only been discussed as hypotheticals until now. The whole idea of using the genetic script to solve a diagnostic mystery exists entirely in the realm of theory.

Still, two days later, on a Saturday morning at the end of June, when his children are at piano lessons and the Whitefish Bay house is quiet, Mayer sits down with his laptop and begins to compose the email that represents Nic's last chance.

"Dear Howard," he writes, "I hope you are well."

As is his nature, Mayer sizes up the person he is addressing and gives careful thought to each word of the email. Since moving to Milwaukee, the doctor has come to know Howard Jacob better. They just missed each other at Massachusetts General Hospital, but when Mayer arrived at the Medical College of Wisconsin, Jacob helped him get his lab up and running. They are on a first-name basis. Now Mayer tries to peer inside the mind of his fellow scientist, an ambitious and competitive man, a risk taker. He knows Jacob has been talking about bringing sequencing to the hospital by 2014. But 2014 is more than five years away, and that's too far off for Nic.

"I'm writing to get your thoughts on a patient of mine . . . that might benefit from a sequencing of his genome," Mayer types.

"This is a unique situation—one that comes along only every

few years. This patient is very ill and has been at the hospital since January. Yet he might be curable if we can make a molecular diagnosis of his illness."

Mayer has been shaping the argument in his mind, given all he knows about Jacob. He is holding out to him the chance to take the first step into the future of medicine and maybe save the life of a very sick little boy. In three paragraphs he summarizes almost three years of fruitless searching and brings Jacob up to date on the current dilemma: a bone marrow transplant appears to be the child's only hope, but without a diagnosis, the procedure will never receive approval.

Then Mayer makes his pitch: "I am writing to ask if there is some way we can get his genome sequenced. There is a good chance Nicholas has a genetic defect, and it is likely to be a new disease. Furthermore, a diagnosis could save his life and truly showcase personalized genomic medicine."

In Mayer's view, the email's encoded message is this: I have a kid who needs this technology, and if you provide it, you will get the credit for saving him. This is what you have been working toward all along. You will get the credit for bringing genomic medicine into the hospital.

Mayer goes on to suggest that "if money is the only barrier, there may be ways to raise funds." Perhaps the most brilliant stroke, however, comes at the end of the email. Mayer encloses a link to Amylynne's online journal and urges Jacob to take a look. "You will get a good flavor for the desperation of this situation."

Mayer rereads and polishes his email until he is certain he has made his best case. Then he hits send.

Chapter 14

Why Are We Here?

JULY–AUGUST 2009

The email gets Howard Jacob's attention. He knows Alan
Mayer, though not well. He knows Mayer as a gastroenterologist and a geneticist who works with zebrafish in his research. Now, as he reads through the doctor's message, he realizes
instantly that this is a simple request to do what he and his team
have wanted to do all along. Mayer has come to him with an invita-
tion to attempt personalized medicine. This is no plan, no research
proposal. This is a doctor with a patient who has exhausted the
existing options. A patient whose last hope is the technology Jacob
has invested his entire career in.

Jacob has never heard of Nic, which is not surprising. Hospitals
have strict confidentiality rules. He will soon learn, however, that
when medical staff talk, they sometimes refer to Nic as "the sickest
kid at Children's Hospital."

Jacob follows Mayer's suggestion. He reads Amylynne's online journal, and, as he does, he thinks of his own two daughters. The story moves him.

"Oh, man, how do you not have tremendous empathy for that family and that child?" he recalls. "Anybody who's a parent, and even if you're not a parent, you just read that and it just tugs at your humanity. This is when you go home, and you just hug your kids and say 'thank you.'"

Clearly, the situation is desperate. On page after page Amy-lynne describes each new attempt at diagnosis. Through all of her son's suffering, there is always the hope that some doctor will figure out what drives this mysterious disease and find a way to defeat it. Each time, the search comes up empty. Somehow the hope remains.

This is a case Jacob cannot ignore. He shares his parents' Roman Catholic faith and carries with him many of the lessons they imparted about one's duty to others. For him, the decision boils down to which is more important: his own risk of failure, or the last chance to save the boy's life.

"Did he have a choice?" Dick Jacob, his father, will later ask. The answer is clear to the geneticist.

Jacob tries to imagine what the Volkers are going through. His own daughters are fourteen and ten, not so far from Nic's age. He and Lisa have been so blessed that their girls are healthy. He thinks what it would be like to watch one of his children wheeled into an operating room, passing into a place where he can neither follow nor provide protection. He thinks how awful that one moment would feel. Then he tries to imagine experiencing that moment more than a hundred times, as the Volkers have. How do they hold on from one day to the next?

But that is the emotional side of the equation, and Jacob is not a man who allows emotion to dictate his actions. The decision really comes down to a single question, one that Jacob keeps turning over in his mind: Is Mayer's request possible?

The kid is dying, Jacob thinks. *Can we do this? Can we read his genome and figure out what's wrong?*

Sequencing the young boy's genes would involve departing from an intricate and well-thought-out plan. It would mean bringing a new technology to patients a full five years ahead of schedule.

Jacob has nowhere near the resources of the big DNA-sequencing centers: neither the staff nor the machines. His advantage is a hand-picked team that believes in, and is working on, the mission of bringing sequencing to patients—a team that is, as they like to joke, drinking Jacob's Kool-Aid. His vision is what has brought them all here to the Medical College.

Now Mayer has forced him into a decision.

"I had been going around saying we should be doing this," Jacob says, "and Alan Mayer basically called my bluff."

Jacob needs data to make the decision, and members of his team should have it. He walks down the hall to find Joe Lazar.

Lazar, a molecular biologist, did research into mad cow disease at the University of Maryland before Jacob recruited him to the Medical College in 2002. Jacob has come to trust Lazar, who is nearly a decade older than he is. He has given Lazar more and more responsibility, and they've become close friends. Lazar's boss in Baltimore had offered to double his salary if he stayed, but Lazar chose Jacob and his vision. He wanted to play a role in bringing DNA sequencing to actual patients, and he believed that Jacob could get it done.

Over the years, his talks with Jacob have given Lazar one overriding impression: "This guy is crazy. If he gets something in his head, he will do it."

Jacob stands in Lazar's doorway. He looks excited. He tells Lazar of the sick little boy over at Children's, describes how sequencing might save his life. "This is our chance," he tells Lazar.

Jacob is already making checklists in his head. They will need someone to generate the data and someone to analyze it. He forwards Mayer's email to members of his team. Elizabeth Worthey has already read it when her boss appears at her door.

"What do you think?" Jacob asks.

Jacob hired Worthey, a Scottish data expert, a little more than a year before from Seattle Biomedical Research Institute. There, she had been focusing on the genomics of diseases found in the developing world: malaria, tuberculosis, and leishmaniasis. At the job interview, Jacob had impressed and inspired her with his plan to begin sequencing the DNA of sick children by 2014. What Worthey remembered most was Jacob's eloquence as he laid out his vision of genomic medicine. And since arriving in Wisconsin, she has watched the Medical College purchase the 454 Life Sciences sequencing machine, the expensive, high-powered tool that confirms Jacob's 2014 plan is more than just talk.

Now he is at the door to her windowless office raising the possibility of sequencing their first patient a full five years ahead of schedule. That morning Worthey has been examining data from the new 454 machine. It is their first set of human genome data, the genetic scripts of a family whose child has died of a cardiovascular disorder. The project is research only, a way to test the protocols they will employ when sequencing is used in the hospital.

But Jacob is telling Worthey that the moment has arrived. Here is a patient whose fate could hinge on having his genes read, whose illness has defeated an army of doctors.

The research and testing of clinical sequencing have not been done. The protocols have not been written. None of that matters to Nic Volker's family. They understand that Nic's doctors have run out of proven tests and treatments. This untested step into the realm of the genome—this is it for Nic. This is his last chance.

To Worthey, there is nothing difficult about the question of whether or not to sequence the boy. There are no other options.

Yes, she tells Jacob. We should examine the boy's genes. That's the best and only chance of discovering what is causing his disease.

Sequencing is not a cure, of course. They might not find a genetic cause. Or, more likely, in Worthey's view, they may find the cause and discover they have no way to treat it. But weighing against this risk is the certainty that if they do not try, they cannot reach a diagnosis. Without a diagnosis, Nic will die.

Given those facts, Worthey is eager to get started.

As she contemplates the sequencing of Nic Volker, she knows it will differ from anything she's ever done. She was hired as part of the Medical College's rat genome database team, and the vast majority of her work has been research. She has grown accustomed to thinking that what she does might, years in the future, help thousands of people. She does not know any of those people. There is no face to think of for inspiration as she works, and it might be years before she sees the results of her labor.

But this will be different. This time, she will be trying to help a single person. And if she is to succeed, her work must proceed much faster.

There is a major difference between the data that Worthey is examining on the morning Jacob appears at her door and the data that is likely to come from this little boy. With the research project, Worthey does not face the same pressure. It is too late to save a life. She can only learn from the script.

Learning something from sequencing Nic won't be enough. This time, the primary goal will not be to advance science but to rescue a patient.

As Jacob pulls more members of his team into the discussion, few are as optimistic as Worthey. The prevailing view is that once the sequencing machine spits out Nic's genetic script, an endless stream of the letters C, A, G, and T, the scientists will struggle to make sense of it. In all likelihood, they will find thousands and thousands of places where Nic's script differs from what is considered normal. How will they ever figure out which of those differences has caused an intestinal disease that doesn't even appear in the medical literature?

This is, in fact, Jacob's biggest concern—that they will sequence the boy's genome, then find themselves unable to do the right analysis. Even in its research, the team has not analyzed many genomes.

Alan Mayer keeps the pressure on. A week after sending the email, he calls Jacob.

"Can you do it?" he asks.

"Theoretically," Jacob replies.

"No, can you do it?"

To answer that question, Jacob turns to Worthey and Michael Tschannen, the research associate who runs the college's sequencing lab. He calls both into his office. How confident are they that they can do the sequencing and analysis? he asks. He tells them to consider the question and get back to him quickly.

Beyond what she calls "my naive optimism," Worthey sees cause for hope. The very things that make Nic's illness so mysterious and daunting—its early onset and fundamental strangeness—make her confident that the cause must be genetic. It is just so strange, so unusual, that she believes something will stand out among the 3 billion letters of Nic's script.

It will stand out because it will be different from the yardstick geneticists use to measure what is considered normal, the reference genome.

Though some aspects of the Human Genome Project were more esoteric and harder to explain to the public (the need to produce genetic scripts for *E. coli*, yeast, fruit fly, worm, and mouse), the reference genome offered a more straightforward concept: figure out what a typical human genome is, then you can compare the scripts of people with strange illnesses and look for differences. That yardstick was the ultimate goal of the project, and it was a massive undertaking. The first human genome took hundreds of machines and seven years to assemble. The cost of the human work alone was about $600 million. The overall project included work on yeast and other organisms and cost $2.7 billion.

Although there was a suggestion that the project, launched in 1990, had finished when the White House held a ceremony on June 26, 2000, only a rough draft of the genome had actually been completed. Craig Venter and Celera Genomics had challenged the government-funded enterprise with fast-paced sequencing of their own. So when President Bill Clinton honored the accomplishment at the White House, it was largely a ceremonial sharing of credit between the public and private projects. In the ensuing years, the public effort moved forward, quietly producing a finished version in 2003—the 3.2-billion-letter signature of a human being.

But whose signature was it? What did it mean? Those questions would assume importance as the scientists and doctors in Wisconsin tried to apply the knowledge to Nic Volker's baffling illness.

The first human genome, now known as the reference genome, represented not one but twenty-eight people. Yet, remarkably, most of the genome came from one person, according to Lisa Brooks, director of the genetic variation program at the National Human Genome Research Institute. About 70 percent came from RPC 11, an African American man (RPC stood for Roswell Park Cancer Institute in Buffalo, New York, one of the centers that collected samples for the project). The other 30 percent of the reference came from the genetic scripts of the remaining twenty-seven people, all of them American, though of differing ethnic backgrounds. All had responded to an ad seeking volunteers. Together they formed a genetically diverse cross section of humanity. Some of the volunteers gave blood, some sperm, to be sure script from the Y chromosome was captured.

The private sequencing effort led by Venter and Celera produced its own reference genome drawn from just five people, including Venter and his colleague Hamilton Smith. However, in the years following the joint announcement that the public and private sequencing projects had succeeded, the Celera results were less often used as a yardstick. When scientists spoke of the reference genome, they usually referred to the results of the public effort.

The reference genome was not perfect. How could scientists suggest what normal might be on the basis of twenty-eight humans out of 6 billion? How could they know what was rare and what was common? They were not even certain what many genes did or even how many genes a person had. A good way to start an argument at a scientific conference was to ask any two geneticists for the total number of genes in the human body. Some would say 20,000. Some 25,000. Some far more. Most would couch their estimates with the phrase "It all depends." It depends, that is, upon how one defines a gene; there are RNA, or ribonucleic acid, molecules that are considered to be genes by some scientists but not by others.

Another point to consider when constructing a reference for normal was that all humans have very rare variations in the script. In other words, it is normal to be abnormal in thousands of places in our genome. And some of those variations almost certainly have found their way into the yardstick.

Yet, imperfect as it might be, the reference genome was a starting point. Doctors and scientists could begin with the mind-set that since those twenty-eight people had no known rare diseases, the genetic script drawn from them illustrated what works for us as human beings. Some differences produce safe, acceptable variations, making a person unusually tall or red-haired or blue-eyed. Other differences in the genome do not work; they prevent the proper manufacture of a key protein, causing a serious breakdown in the human machine. Proteins are crucial to everything humans do—eating, breathing, even thinking—and many diseases, including cystic fibrosis and sickle-cell anemia, stem from the failure to make a protein properly.

So the reference genome was a useful tool as long as one kept in mind that departures from the reference were to be expected and could have a variety of consequences—good, bad, or indifferent.

Another major caveat had to be considered when searching for normal. Scientists were most familiar with the specific parts of genes that contain recipes for making proteins. Those segments make up just 1.2 percent of the entire genome, a portion called the exome. What much of the other 98.8 percent did amounted to a big question mark at the time doctors were discussing whether or not to search Nic's script. For years the assumption had been that the vast majority of the genome was simply "junk DNA." The notion was comforting, though it also seemed to reflect a certain arrogance—the idea that if scientists hadn't discovered what those parts of the genome did, it must be because they did nothing or at least did nothing important.

The genome was treacherous territory in which to make assumptions; so much of it was poorly understood. Yet in 2009, certain assumptions were a necessity. They were all the scientists delving into Nic's genome had to work with. There was no published record of any doctors sequencing someone's genes and using the information to pinpoint the cause of a disease. In order to use Nic's genetic script to determine a treatment, in order to make sense of the massive amount of information the script contained, Jacob's team would have to rely on assumptions.

Within a day, Worthey and Tschannen are back in Jacob's office. It will be too expensive to sequence every last letter of the boy's DNA, they say.

But there is another possibility, one that will save money and help the scientists focus their search for the cause of Nic's illness. Instead of scanning all 3.2 billion base pairs, Worthey and Tschannen suggest that they could zoom in on the exons, the portions of each gene involved in making proteins. Since protein problems are the cause of many genetic diseases, there is a good chance that the explanation for Nic's disease lies somewhere in his exome. A tool has been invented recently for just this purpose, to allow researchers to narrow their search to a person's exons. The Roche NimbleGen Exome Capture Array, often referred to as the exome chip, was developed just a few miles from Nic's home. The chip captures only the exon portion of each gene. If they were to use it, they could cut the total cost from $2 million to perhaps less than $100,000.

Are they sure they can do it? Jacob presses.

Worthey and Tschannen do not hesitate: Yes.

Then it's time to hold a larger meeting, Jacob says.

As Jacob emails and raps at the doors of the other doctors and scientists, he sees that despite some initial reservations, there is a strong consensus for going forward and plunging into their first case of genomic medicine.

Within days of his second talk with Worthey and Tschannen, Jacob calls a meeting of ten doctors and scientists. He is confident they will support sequencing Nic. After all, he hired many of them with the idea, even the promise, that he would lead them to this moment.

If the team hesitates, Jacob is prepared to bring more people into the discussion, to figure out how they can proceed with sequencing. But he does not admit this to his colleagues.

In the conference room that afternoon, there is pushback from some, a concern that they will ruffle feathers by leaping into something costly that has yet to prove its worth in a patient. And how will they face the boy's parents if they find the problem and realize they are helpless to fix it?

Among those with serious reservations is James Verbsky. He wishes he could share Worthey's optimism. The immune specialist has been among those treating Nic, trying to pierce the mystery of his disease.

Verbsky and his peers have suspected for some time that a problem with Nic's immune system is causing the boy's illness. But the fact that it does not fit any of the diseases they've looked at makes them worry. What if the problem is not with the immune system at all?

Only weeks ago, when his last promising lead failed to pan out, Verbsky had given up seeking an answer to the disease. He hated to give up. Yet, even though he has exhausted every idea he can think of, he cannot embrace this new one. Sequencing all of the young boy's genes is beyond a long shot. When Verbsky ponders the idea, his bottom line is this: *They will come up with 20,000 possible mutations. Then they'll come back to me wanting to know which one. Which is the culprit?*

Any difference in the script could turn out to be a mutation; a mutation is simply a variation that is harmful. The immunologist knows that even though there are only about 21,000 genes in the human body, a person may have multiple differences within a single gene. He believes that to go through all of the variations in search of the one that is killing Nic will take a lifetime. It cannot be done.

He is not silent in his skepticism. "I laughed," he recalled. "I said, 'Go crazy. I hope you find it.'"

In truth, the idea of using sequencing cuts to the heart of why Verbsky works in pediatrics. He likes the ethical simplicity, the idea that you do anything to save a child. By this measure, he should have little difficulty with the decision to scan Nic's genetic script.

But Verbsky believes that sequencing carries a risk. While scanning thousands of variations in Nic's script, they may find one that looks like a match for the child's symptoms. Maybe it is a variation in a gene linked to the immune system or associated with wound healing. They all might think the variation looks like the cause, must be the cause. Based on all of their knowledge and instinct, the doctors give the go-ahead for Nic to receive a bone marrow transplant.

Only they turn out to be wrong.

The variation that looked so promising isn't the one they've been hunting. Something else has been causing Nic's disease, something that couldn't be fixed by a risky bone marrow transplant. But because doctors sequenced Nic, because they thought they had plucked the answer from thousands of variations, they green-lighted the transplant. And in the worst of all possible scenarios, it is the transplant itself, not the disease, that kills Nic. Instead of saving a boy and ushering in a new era, the doctors end up violating the most basic principle in medicine: first do no harm. That is the risk they take in attempting to use Nic's genetic script to solve the riddle of his disease.

Verbsky is not alone in doubting that sequencing will work, but he is alone in focusing on such a fearful outcome.

Jacob listens to the concerns voiced by Verbsky and others. Then he impresses upon the scientists and doctors in the conference room one simple fact: if they do not attempt to find a genetic cause for Nic's illness, they are leaving the boy to die.

He allows the sense of urgency to settle over the room.

"Let's give it a try," says David Dimmock, a pediatric genetics specialist. Dimmock is, by nature, a pessimist. In his experience, the very best genetic tests that doctors use result in a diagnosis only about 10 percent to 20 percent of the time, and he cannot imagine that sequencing will offer better odds. Dimmock doubts it will work, but he and many of the other doubters in the room have come around to the idea that it is better to go forward than to leave Nic and his family without hope.

Tschannen is among those who are skeptical they will succeed, but he has a feeling this is what his career has been building toward. He remembers growing up in Grayslake, Illinois, just a few miles from Six Flags Great America amusement park, yet utterly apart from it when it came to childhood pursuits. There was a wooded area near his home where he captured frogs and built traps to catch bass. He grew to admire the way his father, an earth science major who sold water treatment supplies, could hold up a rock and say what it was called and how old it was.

Tschannen studied biology in college, hoping to work in the state fish and wildlife office. What he did not want to do was work in a lab—which was, of course, what he wound up doing.

Since coming to the Medical College in May 2000, he has immersed himself in the effort to learn from the rat genome. Like the others, he has been sold on Jacob's vision of bringing genomics into the hospital. For eight years he has worked with rats and come to view genetics not as voodoo but as something you can see. You can genetically engineer a rat to have heart disease, then watch as the animal grows up and develops the disease.

In 2008, Jacob told Tschannen that he was going to have an important role in ordering, installing, and operating a new high-speed sequencer, perhaps a turning point in the plan to translate the human genome into medicine.

Now Tschannen sits in the middle of a meeting he had not expected so soon. And even though he knows the plan was for 2014, and even though he knows there is a strong likelihood of failure, he feels they are compelled to go forward. He speaks up, repeating a question he raised in one of the earlier meetings with Jacob and Worthey. "If we choose not to do this, why are we here?" he asks the room. "This is the entire reason we're here."

There is no need for a vote.

Jacob knows, everyone in the room knows: they are embarking on not just a risky project but a costly one. Money is going to be a sticky issue.

At this point, it is far from clear what it will cost to sequence a patient's entire genetic script, all 3.2 billion base pairs, and analyze the results. A very rough estimate puts the price at about $2 million. That kind of money represents the outer limit of what a hospital might spend on some of the most expensive accepted treatments—care of an extremely premature baby born at twenty-two weeks and hospitalized for several months in intensive care. Hospitals do not spend such sums on an experimental treatment for a single patient.

Fortunately, they can use the idea suggested by Tschannen and Worthey, that of sequencing the exome. Though they will still have to find significant funds, the cost will be more manageable.

But Jacob decides to address the money issue on another day. The consensus among the ten doctors and scientists meeting that day in the summer of 2009 is to proceed with sequencing Nic, to scour his exons and take the chance that it is the protein-producing part of his genes that contains the cause of his illness.

Leaving the conference room, Worthey feels they will succeed. They will sequence Nic Volker's 21,000 or so genes. Then they will find the culprit behind all of the havoc in Nic's body, and if they can, they will save him.

Tschannen feels pleased, too, but anxious. He has wanted to take on an important project, but this carries more pressure and responsibility than he'd ever expected.

One day, not long after the meeting, Jacob leaves the Medical College for a five-minute walk across the campus to Froedtert Hospital. Tschannen accompanies him, sharing his concerns about the sequencing project. They walk and talk until Jacob pauses.

"If we do it and it doesn't work, it will be big," he tells Tschannen. "If we do it and it does work, it will be monumental."

Jacob resumes his walk. He explains that until now, everyone in the scientific community has been talking about attempting this sort of thing. No one has actually done it. "We have a patient in hand," he says. "No one else is going to be doing it this year, or maybe in the next couple of years."

They arrive at Froedtert, and Jacob walks inside. Tschannen turns back toward the Medical College.

They are committed.

Chapter 15

Unknown Territory

JULY 2009

No one will read Nic's genetic script unless the Volkers allow it. So they have a decision to make.

In one sense, the family's choice seems obvious. The doctors have told them that they've tried everything else, except a bone marrow transplant, and they won't perform the transplant without a diagnosis. Sequencing Nic's genes appears the only way to end the stalemate. There are no guarantees that the boy's genome will yield an answer, but it's the last hope they have. To say no would be to doom him.

In another sense, however, deciding to open the child's genome to scientific inspection is not a simple matter, not to be taken lightly. Scientists have acknowledged from the start that the potential benefits of reading the human blueprint come with potential perils. The genome is a Pandora's box. What lies inside is as private as information gets in this era of National Security Agency snooping, computer hacking, and identity theft. We don't want others knowing what's in that box. We're not even sure we want to know ourselves. Peer inside the genome, and what's seen can never be unseen. A glimpse of our future. Our health. Our life and death.

In Greek mythology Pandora was given a jar as a wedding gift and warned never to look inside. Although the jar became a "box" over the years, the symbolism is the same: Pandora's box was an object of overwhelming curiosity; yet to open it was to release unintended and terrible consequences.

Today the genome evokes a similar aura of curiosity and fear. How badly do we want to know what's inside, and how much are we willing to risk for that knowledge?

For one thing, the decision to read our genetic script goes beyond a personal choice. Because we get our genes from our parents and share many of them with our brothers and sisters, anything we learn has implications for others. So looking at Nic's script also means revealing secrets embedded in the genes of his parents, siblings, and other relatives.

By its nature, the genome is the broadest imaginable search of a person's being; the exome, while less broad, still touches on every gene in the body. The advantage of such wide approaches is that they cover sections of chromosomes that doctors would never have thought to investigate. The culprit may be a gene that even the experts would never have suspected. The disadvantage is that doctors see far beyond the illness they're trying to understand. Though they may set out looking for one specific mistake in the script, they're certain to find many more. Any human's genetic code contains thousands of differences from the reference genome, though only some are harmful.

Just how harmful could a difference in the genetic script be? The doctors searching for the cause of Nic's intestinal disease might discover, for example, that he has a mutation that causes Parkinson's disease. Parkinson's symptoms can begin subtly, with an unsteady walk or shaking. What follows is a steady, irreversible decline. Eventually patients lose coordination, and some descend into dementia. There is no cure. Some people want to know if they have the disease in order to make the most of their lives before the decline begins. Others would prefer to live without that knowledge hanging over them. And if Nic has Parkinson's, there's a good chance that a parent or sibling will have it, too. The information in Nic's script could cause a ripple of revelations that would leave his entire family facing difficult decisions. The Volkers and the doctors could face equally wrenching decisions about Nic. If, while searching for the cause of the child's intestinal disease, doctors discover that his life could be cut short by something quite different, should they tell the family? Would Nic have the right to know even though he is far too young to understand the implications?

Even if doctors find only the cause of his intestinal illness and no more in his genes, that information alone holds powerful consequences. Should Amylynne and Sean agree to have their scripts examined, too, thus revealing which one of them gave Nic the deadly mutation? Are they ready to deal with the feelings of guilt and blame that may arise from that knowledge? Are they ready to deal with what the information in Nic's genes could mean for any future children they might have together? For their grandchildren and great-grandchildren?

This is the box the Volkers will be asked to open.

The hope is tantalizing. The genomic landscape is dotted with several thousand diseases caused by defects in a single gene, including Huntington's, Tay-Sachs, and phenylketonuria (PKU). For those diseases, examining the genetic script can provide doctors with a quick confirmation, a way to tell parents definitively what is wrong. Nic's doctors don't know whether his intestinal problems are caused by a single-gene disease, but that is Mayer's hunch and, in all likelihood, the best-case scenario.

Mayer knows that opening the genomic box may not clarify

things. Given our fragmentary knowledge of genes—the fact that we don't know what many of them do or even how many there are—the picture they reveal of Nic's disease may still be murky. Sometimes the script tells a story that's more about risks and probabilities than certainties. For complex diseases, scientists are finding that a mutation or set of mutations may simply raise a person's risk. It's not clear how a doctor ought to translate a 15 percent higher risk of heart disease into useful instructions for the patient. *Try exercising one extra day a week? Maybe resist that third slice of pizza?*

For the Volkers, such considerations are rendered trivial by Nic's condition.

"We need urgent prayers," Amylynne writes in her CaringBridge journal, "now please."

It is July 14 when she writes this. Two of the doctors, David Margolis, the bone marrow transplant specialist, and James Casper, a pediatric hematologist-oncologist, are recommending that Nic be given high doses of chemotherapy. This treatment has proved successful in children with Crohn's disease, and the doctors think it is worth trying, even though every other Crohn's treatment has thus far failed to help Nic.

The theory is that if he survives the chemotherapy—and the doctors believe he will—the drugs should wipe out his immune system, allowing it to reset itself and develop normally.

Amylynne cannot get past that "if." Already she has been warned at least half a dozen times that Nic might not live through the night. She finds it hard to ward off her fears. And now doctors are telling her she must decide on the chemotherapy plan soon, before Nic becomes too sick for treatment.

"What do I choose?" she writes in her journal. "How do I make the right decision? What if I wait? What if I don't wait? Either way I could compromise my son's life."

Such is Nic's condition when Mayer raises the idea of sequencing his DNA. He explains the risks and uncertainties to Amylynne. The information in her son's script may be painful or frightening. There may be consequences for her, for Sean, for the girls.

Amylynne listens, processes—but she does not hesitate. How can she worry about far-off consequences when right now, in the present, Nic is dying? Nic is what matters. She will do anything she can to save him, and Sean feels the same.

"The Docs are talking about wanting to do some high-tech DNA sequencing," Amylynne writes in her journal, "but it will take a lot of paperwork and they still need to find both the facility to do it and the funding."

After years of debating the deep questions associated with the genome and wondering how a family might navigate such difficult territory, scientists are about to find out. They have worried about our privacy. They have agonized over the possibility that our genetic information might be used against us by employers, insurance companies, or computer hackers. They have debated whether most of us are even emotionally equipped to handle the information in our genome. When the doctor describes our "risk of disease" will we instead hear "certainty" and panic?

Doctors who have dealt with families like the Volkers, families on long, frustrating medical odysseys, do not share these fears.

"One of the general attributes of families like that is that they are amazingly resilient," explains Les Biesecker, the chief of clinical genomics at the National Human Genome Research Institute. "Families that go through these experiences are tough as nails because they have to be. They tell us, 'We've been through hell. What makes you think we can't deal with this, too?'"

Amylynne could have told the scientists what happens when fears and ethical considerations are set against the life of your child. When the discussion of a possible bone marrow transplant first surfaced, she talked with Sean about having a baby to serve as Nic's donor, as no one in their blended family is a match. There was no question of ethics, as she saw it, only the matter of saving Nic's life. She was prepared to go through with the plan and have a baby to serve as Nic's donor; it was her pastor who dissuaded her. The pastor explained that she would only be forcing a solution; she must allow God to provide one.

Nic's plight has taught Amylynne how much faith she has.

She does not fear what her family may learn from Nic's genetic script. When the time comes to give final approval to sequencing, she faces a large stack of medical releases that must be signed before the procedure can take place. There are many, many pages. Amylynne does not read most of them. She simply signs her name.

Chapter 16

I'm Going to Tell You a Story

AUGUST 2009

Although he was one of the founders of PhysioGenix, this will be Howard Jacob's last board meeting. Over the past three months, he has sold enough of his ownership stake that he will no longer have a say in how the company is run.

Brian Curry, who is taking over as PhysioGenix's chief executive officer and president, worries that the meeting will be "horrible." He knows of one other instance in which a founder's ownership was diluted and by all accounts, it wasn't pretty. After three months of negotiating, Curry still wonders whether Jacob is ready to walk away from this company he founded more than a decade earlier. When Jacob started PhysioGenix in 1997, it was to make use of the expertise he and his colleagues had built up through their work with rats. The company began selling animals—"rat packs," former CEO John Seman called them—that were bred to have certain traits, such as diabetes. The traits allowed researchers to use the rats for specific experiments. But PhysioGenix has evolved away from that business model and is now finding success using its unique rats to conduct research studies for scientists and industrial customers.

Jacob is actually pleased to be stepping aside; within a year he will be out of the company completely. In his mind, PhysioGenix was based on a great idea but started too early. If zinc finger technology, which made gene editing much faster and cheaper, had been there from the beginning, the company might have been more successful.

Instead of turning tense, the meeting goes well. It is pleasant, no disruptions. Curry breathes a sigh of relief.

When it is clear there is no more company business, Jacob speaks up.

"I'm going to tell you a story, and if you stay, you are going to write a check." No one moves, and Jacob tells them about the young boy at Children's Hospital with a condition that has stumped every doctor he's met.

"I think we can sequence his genes, and I think we can save his life," he says. "Can you help make this happen?"

It takes all of two minutes. Jacob is low key but extremely confident.

Curry is stunned. Not by the fact that Jacob is talking about sequencing all of the boy's genes. Jacob has been saying for years that his goal is to bring this technology to humans. Curry is stunned that after three months of negotiations that are going to take his company away from him, Jacob is already moving on.

"We had made the decision about it," Jacob later said. "It was over. And the Nic thing was the next thing."

After hearing Jacob's pitch, Jeffrey Harris is one of the board members who decides to contribute. Harris has invested in many young companies, including PhysioGenix, since leaving his job as general counsel of Apogent Technologies. This is the first time he is making a donation to Children's Hospital. "Knowing a lot about what Howard is doing in terms of genetic research, I thought, *Wouldn't it be a shame if we couldn't combine the people we have and the technologies we have locally to help this child,*" he said.

Jacob doesn't stop with the PhysioGenix board. The economy is in a recession and funds are tight, but he knows where to turn. He tells Nic's story to business contacts he's made over the years, to colleagues, even to his parents. Dick and Janet Jacob donate $300, which brings another $300 in matching funds from Dick's employer, Abbott Laboratories.

Mike Tschannen and Liz Worthey, meanwhile, call Jolene Osterberger, their sales rep at Roche, the company that makes their sequencing machine. The researchers are hoping that Roche will partner with them to sequence Nic. Such partnerships are common practice in the industry: researchers get access to a company's scientific staff and products, and the company hopes to get mentioned in scholarly papers and speeches. The higher profile the researchers, the more marketing punch the arrangement can pack.

Tschannen and Worthey read from the letter Alan Mayer sent to Jacob. Worthey talks in a motherly tone, explaining the urgency of the situation, and Osterberger thinks of her own three children. Later Osterberger recalls being struck by one thought, nothing complicated: *This is research that could save the little guy's life.* She promises to take the request to her bosses.

Getting Roche's assistance and raising additional money are, in fact, the only way to make the sequencing happen. Nic has exhausted his $2 million lifetime insurance coverage. Even if he hadn't, there would be little chance of getting an insurance company to pay for a costly and untested shot in the dark. Jacob emails top executives at the Medical College and Children's Hospital to ask if they can find funds for the project. They all say there is no money for it. Jacob notes their silence on the issue of whether the project should proceed. It is permission, as far as he is concerned.

"Failure to say 'no' with me is taken as agreement," Jacob said.

Still, he and the other doctors and scientists caring for Nic find themselves in a kind of ethical no-man's-land that lies between research and the practice of medicine.

Jacob must delicately navigate Children's Hospital policy. There is a rule prohibiting staff from raising money to pay for the care of a single patient. So Jacob says that's not what he's doing; He is running a pilot program to test the use of DNA sequencing and its potential to help patients. In other words, he is engaged in research.

But if sequencing Nic is to be research, it must come before the hospital's institutional review board, the body that oversees all studies involving human subjects. Jacob asks the review board for an approval or a waiver, knowing that either will allow him to proceed. J. Paul Scott, the board's chairman, says that reading Nic's genetic script isn't research at all. What the doctors and scientists are doing with Nic is simply the practice of medicine. It's what hospitals do every day; they try to save lives any way they can. Scott grants a waiver.

In truth, sequencing Nic is both research and medicine. It is the practice of medicine, because everything else has been tried and no treatment found to halt Nic's disease. Yet it is impossible not to see that this is a huge step forward in research, one of the first uses of a technology with potential to transform our approach to disease. Although Nic's illness is extremely rare, in fact the only recorded case doctors can find, sequencing offers a strategy that could revolutionize the treatment of many other rare illnesses. All told, rare diseases cut a vast swath of misery, afflicting about one in ten Americans. Most of those rare illnesses are hereditary; their secrets reside in our genes.

Roche, which has been building its sequencing business with acquisitions of companies such as 454 Life Sciences and NimbleGen, has much at stake in the decision to participate in Nic's sequencing. A large corporation must be careful about reaching too far, particularly in the United States, where regulators at the Food and Drug Administration carefully watch what crosses from research into clinical practice. The exome chip Tschannen and his colleagues are using is approved for research only; the FDA has not approved it for patients. Companies such as Roche are wary of violating the agency's rules, and with good reason. The FDA had met in July with 23andMe, a Silicon Valley maker of genetic tests for consumers, to point out that the company was marketing its tests without approval. That meeting would eventually lead to a 2013 FDA order that 23andMe halt sales of its main product, which provided customers with insights into their genetic predispositions to a host of diseases and conditions.

Osterberger gets off the phone call with Worthey and Tschannen believing the effort to sequence Nic would be considered a research project. She immediately calls one boss, then another, to discuss Roche's possible involvement. Within about a week, Osterberger comes back with an answer: Roche will do some of the necessary sequencing.

Jacob's team has managed to walk a tightrope to get the funding and approval to go ahead. Jacob has succeeded in raising about half of the estimated $75,000 they need to read Nic's DNA. Roche's contribution will cover the rest.

Chapter 17

Fine White Threads

A teaspoonful of blood.

That's where the search for a new disease begins. Within a few weeks, scientists hope to drill deep inside the blood to the place where Nic's story lies; not just a story, an entire encyclopedia. Inside millions of pages they hope to find a paragraph, maybe just a word, that explains the dragon Nic has been battling.

To get there, the blood will be reduced to DNA, the DNA to exons, and the exons to the individual letters of the genetic script: A, C, G, and T.

In an era when the mining of our genes has turned increasingly to high-speed machinery, some tasks still call for a human touch. So the teaspoon of Nic's blood does not proceed straight to the next-generation equipment. The sample contains material—red blood cells, platelets, and plasma—that must be removed in order to capture Nic's DNA. Then, to reach what Francis Crick called "the secret of life," scientists must pierce the nuclei of what remains, the white blood cells.

There is a machine designed to extract DNA from blood, but Mike Tschannen, who runs the sequencing lab, finds that the individual strands emerge in better shape when the job is done by hand. He knows right away that the best hands for the job are Gwen Shadley's. A research technologist at the Medical College, Shadley has been processing DNA for more than ten years. She is meticulous, and Tschannen has a deep trust in her. Early in his career, she gave him a tour of the lab, impressing him with her knowledge and her commitment.

Shadley regards her work as a special calling.

The oldest of four girls, she grew up on an eighty-acre farm in the small city of Bryan, Ohio. The family raised a few beef cattle, dairy cows, and some sheep, and there was endless work for the girls. Shadley cultivated and harvested wheat, corn, and soybeans. She milked cows, drove a tractor, made hay, and fed the animals at night. The work was hard, often exhausting, but it was a good life. When the day's chores were done, the sisters would go down to the pond and swim.

Shadley's father was an amateur naturalist who used farm life and landscape to teach his children. He would take his eldest daughter to the woods to cut logs, and he'd point out the different trees. When he butchered cows, he cut out the heart and showed Shadley the different valves. He'd cut out pieces of the lungs and explain how air flowed through them.

Soon her own curiosity took over. One day, Shadley's mother went to the freezer and let out a shriek.

"I don't want to go to the freezer and find eyeballs staring out at me!" she cried. Shadley had cut out a cow's eyeballs and placed them on ice to preserve them.

She was interested in science from an early age. Science explained how everything was put together and provided deep insight

into nature's contradictions. She often thought of a line from Scrip-
ture: "I am fearfully and wonderfully made." For Shadley, there was
no better proof than in the nature that surrounded her. "It was fun
to cut things out and see that," she would say.

Shadley got her undergraduate degree from Biola University,
a private Christian school in Southern California, then returned
to Ohio for a master's degree in developmental genetics from the
University of Toledo. She did much of her experimental work with
fruit flies.

In 1998, she was divorced, living in Milwaukee, and looking
for a job when she heard about a research associate position at the
Medical College. The majority of the work would be extracting
DNA. Blood samples would come in from all over the state for a
project launched by researchers at the school. They were investigating the role of genes in obesity.

Like the chores she'd grown up with on the farm, the tasks in
the lab were not especially difficult and might have seemed boring to someone else, but Shadley found the work extremely interesting. There was a set of routine steps to follow, but there were
also certain tricks that could be applied to coax the DNA from
difficult samples. In one of the final stages, the DNA is supposed
to strand out, but frozen blood samples often proved challenging.
She learned to put glycogen in the sample; the sugar would bind to
the phosphorus backbone of the DNA and pull it out of solution.
She would also place the tube on an electronic rocker to move the
process along. Sometimes she would leave the solution for an hour
or so, sometimes overnight. She learned to adapt her process to the
sample at hand.

Shadley, now in her eleventh year of extracting DNA, believes
the work is not simply a matter of following a recipe. You must
have a passion for it, and she does.

Samples are always anonymous, and the one that Tschannen now brings her is no different. Yet Shadley knows this is no

ordinary vial of blood. For one thing, it reaches her bench less than an hour after it has been drawn from the subject. That tells her the job is urgent.

Tschannen does not tell her where the blood has come from, whether the person is alive or dead. Only this: "Handle it with tender loving care. It's a very special sample, very likely the most important sample of our careers."

He does add one more thing: they need the DNA as soon as possible.

In most cases, Shadley tells people it takes three weeks to get the best results. The longer she has the sample, the more DNA she can pull from it.

Knowing that time is a factor, she tells Tschannen, "Let me allow it to sit overnight so I can get you the most DNA."

Shadley works in an old section of the building in a brightly lit, windowless lab the size of a small living room. There are several lab benches, including a low bench about five feet long where she begins processing the sample. Nearby sits a boom box that she and a colleague sometimes tune to easy-listening music to relax them as they work. Today it is silent. Shadley likes the room fairly quiet when she is processing a sample. When she has many samples, she must take care to ensure that one does not come into contact with another.

The work with Nic's blood progresses smoothly. First Shadley uses a solution to burst the red blood cells. Then she moves to a centrifuge, which spins at great speed to separate out the plasma, platelets, and cell debris. At the bottom of the test tube lie the white blood cells. She adds detergents to break open those cells, releasing the DNA inside. Then she extracts the other cell contents—proteins, sugars, fats. What's left gets poured into a test tube of alcohol solution.

She places the tube on a machine that rocks gently back and forth. Her work has taken about four hours.

Now comes the part that leaves her in awe no matter how many times she does it. Tiny strands of DNA, like fine white threads, drift through the solution and clump together, visible to the naked eye. Those delicate white threads remind her of the farm, of her childhood, of the times she helped ewes struggling to give birth. The strands are life reduced to its essence.

"It's the wonder of life," she says. "That is the very code of somebody's life. Everything that you hold in your hand in that test tube, that is the key to somebody's life."

She lets Nic's sample sit overnight. The next afternoon, about twenty-four hours after she began work on the blood, Shadley presents Tschannen with two tiny vials. They contain Nic's DNA, his genetic secrets reduced to a clear liquid unrecognizable from water. She wonders what is in this sample, whom it belongs to, whether it will help in some way. She knows that the strict patient confidentiality laws prevent Tschannen from telling her those things—if he even knows them himself. So Shadley hands over her work, content in the knowledge that she has done her job as best she can.

Although the Medical College has its own sequencing machine, although Tschannen will be using that machine extensively to analyze Nic's sample, the first run is to be done by Roche, the Swiss biotech giant that has agreed to help.

Tschannen picks up one of the vials of Nic's DNA, removes a large drop of the liquid, and places it into a container surrounded by dry ice. He sends the whole package by overnight mail to Roche.

Roche's 454 Life Sciences division in Branford, Connecticut, performs the first sequencing of Nic's genome. Technicians begin by breaking the long strands of DNA with their 3.2 billion chemical base pairs into shorter, readable stretches. Using pressurized nitrogen gas, they shear Nic's DNA into segments of roughly five hundred to eight hundred bases. Each one of those segments is fitted with a short piece of synthetic DNA, a tag that will be used later in the process.

The segments are loaded onto a special chip the size of a microscope slide that captures only the exons, the portions of the genetic script that contain the recipes for proteins. The chip they use is one conceived of and built by scientists at Roche NimbleGen in Madison, just a few miles from Nic Volker's home. It was introduced earlier in the year, about the time Nic's disease kicked back into gear, forcing him to return to Children's.

Its inventor is Thomas Albert, Roche NimbleGen's senior director of research. As DNA sequencing became faster and more affordable, Albert realized it would supplant many of the single-gene tests that were common at the time. That inspired him to come up with ideas that would "play nice" with DNA sequencing, rather than compete against it.

Companies were developing ways to read the genome more rapidly and less expensively, but the ability to sequence selected parts lagged behind. No one had yet figured out how to isolate the exons. There are some 200,000 to 300,000 exons, each an average of 150 bases long, scattered across the genome's 3 billion or so chemical base pairs. Variations in exons cause proteins to malfunction, which can lead to major diseases. Sickle-cell anemia, Tay-Sachs disease, cystic fibrosis, and many others are caused by mutations in the exons.

NimbleGen's core technology, invented by three well-known University of Wisconsin–Madison researchers, gave the company the ability to quickly design millions of strands of synthetic DNA on a computer. Albert realized that those strands could be used as probes that would recognize the exons and pull them out of the genome. That way, researchers could save time and money by focusing on the part of the genome most likely to be involved in disease.

The probes on Albert's chip capture Nic's exons by taking advantage of a basic feature of DNA: each chemical base has a counterpart that it attracts rather like the way two magnets couple.

Adenine always pairs with thymine; guanine always pairs with cytosine.

Albert's probes are designed to attract and capture only the exons. Once they've done so, the rest of the DNA—the nonexon portion—is washed away. The exons are then transferred from the chip into a solution in a small plastic test tube.

The Roche technicians add tiny beads to the exon solution along with an oil-and-buffer liquid and shake the entire mixture. The shaking creates an emulsion, a little like an oil-and-vinegar salad dressing. Millions of tiny droplets form within the oil; the oil keeps each one separate. Each droplet traps a single bead and a single exon fragment.

Here's where the synthetic DNA tags Roche attached to the segments of Nic's exon come into play. Each tag is attracted to a matching chunk of synthetic DNA on one of the beads. This match allows every bead to grip onto a targeted segment of Nic's exon.

The emulsion also contains an enzyme that makes droplets into ferocious copying machines that replicate Nic's exons. As a result, each bead creates hundreds of thousands of copies of the same exon fragment. Many copies are needed to generate a signal that's strong enough for the sequencing machine to detect.

The beads are spread over a sequencing plate the size of a standard Post-it note. The plate contains millions of tiny wells, each large enough for only one bead. Once all the wells are filled, the entire plate is loaded into the sequencing machine.

The machine reads each segment of Nic's exon, letter by letter. The process takes place at high speed, with the machine reading about a million segments, or roughly 100 million bases, in a single twenty-four-hour run.

One at a time, the machine pumps a stream of each of the four bases. When A is pumped, all of the complementary Ts in Nic's fragments light up. When G is pumped, all of the complementary Cs light up.

A camera photographs the pattern of light flashes, creating a map of the plate, where all the As are, all the Ts, all the Cs, and all of the Gs.

A computer attached to the sequencing machine translates the photos into Nic's sequence. Then, using algorithms based on the reference genome that recognize where the segments are most likely to fit, the machine reassembles all of the fragments into the complete string.

In Wisconsin, Elizabeth Worthey has not been content to wait for the opening deluge of data. If the scientists are not prepared, they will drown in all the clues, the variations between Nic's script and the model developed by the Human Genome Project. Since the second week in July, she and research scientist Stan Laulederkind have been poring over the medical literature, noting every time a gene has been linked to one of Nic's key symptoms. Often papers connect an entire pathway—a group of interacting genes—to a symptom. In those cases, Worthey and Laulederkind include all of the genes mentioned. By August, their master list of potential suspects has grown to more than two thousand. The list includes genes linked to inflammation, such as NOD2, and others that play a role in gut development, such as the GATA genes. The Medical College scientists are hoping that one of those two thousand genes is the one they're hunting.

As they construct the master list, two assumptions guide them. The first is that the difference in Nic's genetic script, the one causing the holes in his intestine and the poor healing of his surgical wound, must cripple an important process in the body; otherwise his illness would not be so severe. The second assumption is that the difference they're looking for is extremely rare, likely undiscovered until now. Although the gene may be on their master list, the specific error in it must be new. The scientists know that because Nic's disease appears nowhere in the medical literature. If they were looking for a mutation that had been found before, another child somewhere in the world would have turned up in a hospital with Nic's symptoms. Those doctors would have faced the same frustrating search that Mayer, Grossman, Verbsky, and the others have been through at Children's. The presumed rarity of Nic's mutation has given some of the doctors—Mayer in particular—the feeling that when they go through Nic's script they will know the right variation when they see it. It will jump out at them.

The master list of suspect genes is only one part of Worthey's preparation for the sequencing results. Facing thousands of possible mutations all over the genome, she knows they need a tool to guide them. For the first time in her career, the task she has been asked to perform is so complex that the crucial tool to accomplish it has yet to be made.

She must make it herself.

Chapter 18

Thousands of Suspects

FALL 2009

I nto a single cell.

Into the dark nucleus.

Into the twenty-three matched pairs of chromosomes. There lies Nic's essence—and his secret.

Scientists are finally ready to put their best technology to the test, to peer into the boy's twenty-one thousand or so genes and read the script. This is the power that next-generation sequencing machines have given the team investigating Nic's disease. It is an awesome, unprecedented power. Yet it is not enough.

To return to the image of a camera scanning the genetic script, scientists are now, in effect, bypassing more than 98 percent of the letters to focus only on the protein-making exome. But that still means the camera must glide over more than 35 million letters. The camera is scanning the genetic landscape for differences between Nic's script and the reference genome, our best approximation of normal.

There will be thousands and thousands of differences. All along, the problem has been that doctors do not understand the nature of the enemy, so they don't know what to look for; they know only that the disease is something they've never seen before. Scientists will face much the same problem when they comb through the thousands of differences between Nic's sequence and the normal sequence: which of all those differences is the mutation, the terrible flaw in his molecular machinery?

As Howard Jacob's team prepares to mine Nic's exome for answers, the boy's doctors take the extreme approach to his disease that has been discussed before, the approach that frightens Amylynne. They destroy her son's immune system using a powerful chemotherapy drug called Cytoxan. Then they allow Nic's stem cells to rebuild his immune system from scratch. The procedure is not the same as the transplant they have talked of giving Nic. A transplant of bone marrow from someone else would give the child an entirely different immune system. This procedure, known formally as immune ablation, is more like rebooting a computer. Still, it is agonizing, and not a step the doctors or Amylynne take lightly. The child's fever hovers around 104 degrees. He vomits as many as twenty times in a single day. One day, he looks into a handheld mirror and sees that all of his hair is gone. He shrugs and looks away.

By early September Nic is recovering, and his immune system has successfully reconstructed itself. He is able to go home. He gets to jump in leaf piles and go trick-or-treating. He can even eat some of the candy. In an effort to keep the disease at bay, the doctor allows him to eat just one new food each week. Mayer tells Amylynne that Nic's disease is in remission, but they both understand what that means; Nic's disease has been in remission before.

Around this time, Liz Worthey begins work on the new program to make sense of all the variations and oddities that the researchers will find in Nic's sequence. Working with Medical College software developers, she begins designing an entirely new tool to assist in the effort to pin down the origin of Nic's disease.

The tool Worthey and her colleagues construct must help them to bypass the vast number of genetic variations that do not have dire consequences.

The scientists will be looking at the exons, the portions of the genome that determine the proteins our bodies need to function properly. As they search through Nic's genes for a mutation capable of causing his terrible disease, the scientists will encounter three types of variations. Some changes in the script still allow our bodies to make the correct amino acid. Other variations alter an amino acid without disrupting the overall chain of amino acids that forms a protein. So even though one of the amino acids is different, we still make the right protein in the right amount—imagine this as the equivalent of rewording a few paragraphs in a book but allowing the meaning to stay the same.

Then there is a third kind of variation, the kind that scientists believe is causing Nic's disease. This kind prevents the manufacture of an important protein and leaves a vital bodily process in disarray. In such cases, the typo results in a more serious problem. Key sections of the book are suddenly missing, or a critical chapter ends midsentence. The mistake omits information the reader needs to make sense of the story.

Scientists keep lists of known genetic variations and how they affect different species. One database called dbSNP describes single-letter variations and what is known about each of them. But single-letter changes are only one type of variation. There are a range of others—for instance, there are portions of the genome where the same letters repeat, perhaps ten times in a row in some people, twenty times in others. Databases likewise exist for those types of variations. Each database is continually updated as scientists discover new variations and new information about known variations. The problem is that there is no master list, no single tool Worthey and her colleagues can consult as they scan Nic's script. They will be forced to plow through a number of different databases and to continually assess the harmfulness of each genetic difference they discover. That will make reading Nic's script into a much longer and costlier task.

What Worthey needs is a single, one-step tool that pulls together all current knowledge of genetic variations, a tool that will eliminate all of the inconsequential differences in Nic's script.

Over the course of several months, Worthey and her colleagues build that tool. They make a software program that combines new and existing algorithms, as well as data about genetic variations that scientists have encountered previously, all in order to flag changes in the script that might mean something. It is a vastly more powerful analytic tool than any previously available to geneticists.

Worthey calls the program Carpe Novo, Latin for "seize the new."

When the results arrive from Roche's first sequencing of Nic's DNA, Worthey has an early version of Carpe Novo already working, though she's still adding new information. The sequencing, too, is just a first draft; it likely contains mistakes and errors that will be corrected in the next few sequencing runs Jacob's lab will conduct. Each new run is like a slide superimposed over the previous slides, adding resolution and depth, bringing the picture into clearer focus. For now Worthey has more than enough to start, and she doesn't waste any time. They need to act fast. The numbers are already daunting.

James Verbsky's pessimistic guess that they could end up looking at twenty thousand genetic differences proves to be close to the actual mark. The Roche sequencing comes back with 16,124 differences between Nic's script and that of the reference genome. That's the size of the haystack they are searching.

Worthey's program will act as a kind of filter, narrowing down the daunting list of 16,124 suspects. Once Carpe Novo has reduced the field of suspects, it will then be up to Worthey and others to act as medical detectives, checking the remaining genetic variations to see whether they can be linked to Nic's symptoms.

Fortunately, Worthey is not alone on the case. She works closely with David Dimmock, the pediatric genetics specialist who joined Children's Hospital and the Medical College in July 2008. For more than a year, Dimmock and Worthey have worked together on research that has used sequencing to understand the genetics of liver failure and the genes that cause mitochondrial disease.

A young, fair-haired Englishman, Dimmock brings to his work a deep compassion and a maturity that comes from years of confronting human suffering. More than a decade earlier, Dimmock was working in a tin-roofed hospital in Uganda caring for patients with malaria, HIV, tuberculosis, cholera, and dysentery. Most were children. Every day two, three, four children would die at the hospital.

After his time in Uganda, Dimmock spent portions of the next three years journeying to Tajikistan and Afghanistan, where he worked with medical teams. Tajikistan was at the tail end of a bloody civil war, and he learned how grateful a doctor can become for the small things: sharp needles, clean gloves, and the ability to drive through the countryside in safety.

He also gained an awareness of the limits a doctor must accept. "There are definitely things you are powerless to do," he says. "That recognition that you can't win every time does change your willingness to recognize when you are in a futile situation. There's nothing quite like the death of a child. And if you're seeing it two or three times a day, that doesn't make it any less horrific."

In other words, he has spent years absorbing the two

opposing principles at play in the treatment of Nic Volker: the knowledge that medical care can reach a point of futility and the knowledge that few things in life are more terrible than the death of a child. This time, in order to save a child, the team is redefining the point of futility, pushing past the threshold of what is possible in medicine.

Once Worthey has put the initial sequencing data through Carpe Novo, she and Dimmock discuss the findings. The two scientists bring different backgrounds and strengths to the task. As a researcher, Worthey has honed her patience and focus, learning to work for long periods on studies that may bear fruit only after many years, affecting thousands of patients she never sees. As a doctor, Dimmock has not confined himself to the slow-moving research world. He sees children come in with terrible genetic conditions. He has some idea of what it means to be the parent of such a child and to need a diagnosis—yesterday. Although their roles blend sometimes, Worthey is the data miner; Dimmock, the clinician. She uses computers to pry information from the millions of letters in Nic's exome. Dimmock tests her insights against Nic's symptoms, looking at how potential mutations might cause the specific problems inside the child's body. After they have combed through the initial results, new data comes in and they plow through it, bent on narrowing down the field of suspect genes.

At the end of September, Mike Tschannen performs three more sequencing runs with Nic's DNA. He works ninety-two hours in eight days. He has taken to heart what Jacob told him: this effort could prove monumental for everyone. Scientifically, what they are doing will likely be counted a success if the equipment works and if they get an accurate picture of Nic's genetic script. If sequencing then allows them to solve the mystery of the boy's disease, it will be an even greater achievement. But Tschannen has developed his own idea about what's at stake. When he thinks of Nic and his family, Tschannen knows they have a very different definition of success from that of the scientific community. The work will not be a success to the Volkers unless it leads, not only to an under-standing of Nic's disease, but to a treatment. Tschannen adopts the family's view: if Nic dies, they have failed.

In early October, he sequences Nic's DNA for the fifth and final time. At this point, each segment has been read an average of thirty-four times, enough to reduce significantly any chance that a mutation has somehow escaped detection.

This last sequencing run allows Worthey and Dimmock to whittle the initial field of more than 16,000 genetic differences down to thirty-two suspects. These are the genes that appear prom-ising. They meet with others involved in the sequencing project in a large conference room at the Medical College and discuss the early results. Dimmock and Worthey describe the thirty-two sus-pects, focusing on two in particular that have piqued their interest. One is a gene called CLECL1, which is involved in regulating the immune system. Worthey had flagged CLECL1 in her previous list of two thousand–plus potential suspects.

The other gene that intrigues Worthey and Dimmock is XIAP. This gene influences the immune system, though it had not even made Worthey's top two thousand list. In the conference room, there is skepticism about the gene, especially among the immu-nologists. If XIAP could cause such havoc, it would have been implicated previously in inflammatory bowel diseases. But the immunologists have pored through the medical literature. XIAP has no such track record.

Worthey continues to update Carpe Novo as the team works, and as she does, the software program provides new insights about the most promising genetic differences. She finds that some of the thirty-two suspects can be eliminated. Among those discarded is the once-promising CLECL1. The new database shows that Nic's CLECL1 variation is actually relatively common. If CLECL1 were causing his disease, doctors would have seen many people with Nic's illness. But all of the searches and conversations with colleagues around the country have failed to find a single case like Nic's.

By eliminating CLECL1 and others, Worthey and Dimmock have now gone from an original list of more than 16,000 differences down to 32, down to 8. There is only one sequencing run remaining when Worthey prepares to meet again with the immunologists, including Verbsky, to take stock. In preparation, she compiles a full report on the final eight suspects.

She thinks of this as her "hot list."

Although the list does not include CLECL1, it includes another intriguing suspect called GSTM1. This gene has passed through all of the different filters Worthey has designed to screen out irrelevant differences in Nic's script. GSTM1 makes a protein that is involved in pumping chemicals through the membranes of cells. This process is critical in helping to clear harmful chemicals from the body. A mutation in this gene could be deadly if, for instance, you ate fish contaminated with mercury and could not flush the mercury from your cells.

There's just one problem: the doctors know Nic isn't sick because of a failure in this process. He is sick because no matter what he eats, something keeps boring holes through his intestine. It is that symptom, linked to the family of illnesses known as inflammatory bowel disease, that brings Worthey back to XIAP.

All along she has had a suspicion that this gene, an immune regulator no one took too seriously, might be the culprit. The fact that XIAP has flown under the radar might be a sign. They are looking for something unusual. Perhaps this gene has not appeared in their literature searches because doctors have found very few people with mutations in the gene. And perhaps they have found few mutations because there is something important about the gene, something so crucial that a mutation is almost impossible to survive. But in order to convince herself, let alone Nic's doctors, she must have more than a hunch.

So she looks more closely at how the change in Nic's XIAP sequence, though small—just a single misplaced letter—would gain significance as it cascades through his body. The first consequence is that this tiny change causes a different amino acid to be made. While amino acids can differ in some places on a protein, the amino acid in this particular position is always the same—in everyone except Nic.

The amino acid is part of a long chain of several hundred amino acids that makes a protein, also known as XIAP. Worthey comes to believe that this one link in the chain, the one position where Nic differs from the norm, is critical to the form and structure of the protein XIAP.

She has looked closely at Nic's XIAP gene. She wants to be confident when she presents her findings to the others. On November 16, 2009, at 8:41 a.m., she emails members of the team working on Nic's sequence. "Good morning. I would like to schedule a meeting to go through the findings from the exomic genome sequencing of Dr. Mayer's patient," she begins. "We have identified a variant in XIAP."

For Mayer and the other doctors, this is the first word that the sequencers may have found an answer in Nic's genetic script. As soon as he receives the email, Mayer begins searching for background on the gene. He finds something interesting. Only a few months ago, Worthey and Stan Laulederkind searched the medical literature and discovered nothing linking XIAP to Nic's symptoms. That's why the gene was not among their two thousand–plus suspects.

But since then, a new paper has appeared. In August, *Proceedings of the National Academy of Sciences* published a paper in which researchers from the University of California, Santa Barbara, linked the XIAP protein—the one affected by Nic's mutation—to a pathway involved in inflammatory bowel disease.

All at once, Mayer sees the significance. Worthey has found an extremely rare change in Nic's XIAP gene. That change alters the protein XIAP. And now, in this paper, scientists are suggesting for the first time that the protein XIAP appears to play a role in diseases similar to Nic's.

Hours after Worthey's email, Mayer sends one of his own to members of the team. He attaches a copy of the published paper and an accompanying review. The paper, he explains, shows that XIAP is important in sensing microbes in the intestine. Our immune system is set up to identify bad microbes in the intestine and eliminate them. Problems in XIAP appear to throw off the immune system's ability to distinguish between good and bad microbes, meaning that the system may attack the intestine when it should not. The gene even appears to have an important structural role, helping to hold together the scaffolding for the protein; if the gene malfunctions, the scaffolding collapses, undermining the protein. Piece by piece they are assembling a case; XIAP's fingerprint is all over Nic's disease.

Worthey continues performing due diligence, going back over Nic's entire exome, running through various scenarios involving the eight suspect genes.

Her confidence grows. She and Dimmock study the effect this change in the XIAP gene would have on the XIAP protein. They find that Nic's unusual sequence results in reduced production of the protein. The boy's white blood cells wind up with just 60 percent of the XIAP protein needed. Dimmock and Worthey now have strong reason to believe that the change in Nic's sequence is preventing his XIAP gene from functioning as it should.

Dimmock still harbors a nagging doubt. He believes the mutation is to blame for the problem with Nic's immune system, the failure of his natural killer cells to perform as they should. But the inflammatory bowel disease, the ulcers and holes in the intestine, might be unrelated. He just can't be sure.

Worthey feels more certain about XIAP. She makes the gene her prime suspect. Because it is her prime suspect, she places the gene last on the list she presents to the immunologists at the meeting. She gets all of her data in order, laying out a clear path that she hopes will lead them to the same conclusion she has reached: it has to be XIAP.

Her decision to place XIAP last on the list, however, has an unforeseen consequence. At the meeting, the immunologists plow through the other suspects so methodically that they run out of time and finish before they have even begun to discuss XIAP. To Worthey's disappointment, she only has time to tell the doctors that her top pick is the one they didn't get to.

Still, her frustration diminishes when Mayer talks with her after the meeting. He has reviewed Worthey's findings and the recent journal paper, and he understands what they mean.

It's XIAP, he tells her.

Chapter 19

The Culprit

For the first time, Nic's disease begins to make sense. If Worthey and Dimmock are right about XIAP, all of the havoc inside Nic's body stems from the tiniest of errors, a single misplaced chemical base among 3.2 billion pairs.

The error lies on the X chromosome, one of two sex-determining chromosomes. Boys have an X and a Y chromosome; girls have two X chromosomes. And one of many genes on the X chromosome is XIAP. The name stands for X-linked inhibitor of apoptosis, a daunting string of scientific jargon that means something quite simple: this gene, when working correctly, helps prevent apoptosis, a process that makes cells die.

In one specific position on the XIAP gene, the bases in the normal reference genome read thymine-guanine-thymine, or T-G-T. Nic's script differs at this position; it reads thymine-adenine-thymine, or T-A-T. The boy has what amounts to a typo, an A instead of a G.

Yet the error ripples outward, producing one consequence after another, altering what happens inside Nic's body. The three bases in that small sequence make an amino acid. That amino acid is the 203rd in a chain of almost five hundred; together they form a single protein, also called XIAP. Because Nic's sequence is different at this one position, he makes a different amino acid, tyrosine, where the reference genome makes cysteine. Because Nic makes the wrong amino acid at position 203 in the chain, he winds up making too little of the XIAP protein. Without enough XIAP protein, a vital role inside the body goes unfulfilled. The protein is not only supposed to prevent cell death; it has a second function—to stop the immune system from attacking the intestines when food enters.

Nic's typo represents the most terrible luck. If Dimmock and Worthey are correct, all of the trips to the operating room, all of the times doctors told his parents he would not survive the night—all of it stems from one misplaced letter. Imagine a single typo in the 55-million-word *Encyclopaedia Britannica* online edition; Nic's error is even smaller.

The scientists have made a case for XIAP, but thus far it is largely circumstantial. The paper in the medical journal only linked the XIAP gene as a whole to inflammatory bowel disease; it did not examine Nic's specific variation of that gene. Before doctors take the next step and look for possible treatments, they need to be certain they have found the right error in the script, the precise flaw that is triggering Nic's disease. They need to understand how the changes in one base, one amino acid, one protein work inside the boy's body. They need to understand, too, why the medical literature contains no other reports of this illness. They are seeking the logic behind the symptoms, why they happen, and why they happen to Nic and no one else.

Worthey searches for the boy's error in other human genomes. If the error shows up, but not the disease, it will signal that the scientists need to keep searching, that they have not found the origin of Nic's illness. The culprit must lie elsewhere in his script.

By 2009, there are only a dozen or so published human genomes, a minuscule cross section of the 7 billion people on earth. Checking the published genomes, Worthey and Dimmock find no one else with Nic's mutation. But his mutation's absence from a dozen genomes is hardly sufficient to make it rare.

So Worthey widens the genetic net, privately approaching scientists at a conference at Cold Spring Harbor Laboratory, the nonprofit research institute where James Watson still works. The other scientists have access to unpublished genomes, information that is closely guarded. These genomes are the basis for important, as-yet-unpublished studies; the scientists do not want to give away what they're working on before they go to print. But Worthey tells them about the sick child with the mysterious disease; she asks only that the scientists check the XIAP sequence on any genomes they can examine. She gives them precise coordinates for the mutation's position and appeals to an impulse she believes all scientists share. "Everybody gets into science because they want to help people," she says. "They want to do things that will help humanity."

One other factor works in her favor as she talks with the other scientists. Most are in their thirties or forties and have small children of their own. They can put themselves in the shoes of the parents desperate to learn what is making their child so sick. In their hearts, Worthey says, most scientists believe that genomic data needs to be shared. The scientists she approaches deliver.

Dimmock contacts still other scientists by phone, asking the same question: Has anyone else come across the young boy's strange sequence? In all, they check roughly 2,200 human genomes. Not a single one shares Nic's error.

But they do not stop at human genomes. Researchers around the world have sequenced a variety of other species, including fruit flies, rats, mice, cows, chickens, and chimpanzees; all make their own versions of the XIAP protein. The genomes of these different species are available to scientists through the National Center for Biotechnology Information, a clearinghouse for molecular biology data that is part of the National Institutes of Health. Worthey and Dimmock check the genomes of every species they can find. None makes the amino acid tyrosine at this position.

It might seem strange to compare Nic to chickens and fruit flies, but science takes a broad view of the genome. All species have a genetic script, and many species share a sizable percentage of genes. Humans and fruit flies, for example, have about 60 percent of their genes in common. By comparing the scripts of different species, researchers can see the differences that nature allows. They can also see the places where nature does not tolerate alteration, where all creatures require the same script. In other words, comparing genomes shows us what works in nature and what doesn't.

As they proceed from species to species, the scientists find nothing that shares Nic's mutation. The absence of his sequence carries an unmistakable message.

"If all of those organisms have cysteine at this position, then clearly it is important because over all that time it has never been allowed to change," Worthey explains. "If it did change, something bad obviously happened to stop that line from evolving any further. So everything has cysteine."

Everything except Nic.

From an evolutionary perspective, life with Nic's sequence has not been sustainable. The sequence equals death.

In fact, the mutation is even more deadly than doctors had suspected. Until now, Worthey, Dimmock, and Mayer have been focused on determining whether Nic's current symptoms, the inflammation and holes in the intestine, might be tied to the XIAP mutation. But, to their surprise, the sequencing has revealed—and a blood test analyzed by an outside lab has confirmed—that the same mutation has given Nic a second, extremely rare disease called XLP. This disease affects boys only. Although XLP is rare, it has the effect of leaving patients unable to fight off one of the most common human viruses, Epstein-Barr, which can cause mononucleosis. In this respect, Nic has been lucky; he has never contracted mononucleosis. But XLP is a biological time bomb. Most boys with the disease die before the age of ten.

Other boys with XLP do not share Nic's mutation; they have contracted the disease through a different mutation that affects the same protein. The only cure for XLP is a bone marrow transplant—the very treatment that Nic's mother, Mayer, and others have been advocating for the mysterious gut disease. Until now they have lacked the diagnosis to justify the transplant; with XLP, they have one.

Sequencing has revealed Nic's condition to be more complex than anyone had imagined. It has also raised the possibility of a treatment. What isn't clear is whether a bone marrow transplant would cure both diseases.

At this point, the Volkers have yet to hear any of the talk about XIAP, the second disease, or the new justification for a bone marrow transplant. To the extent that it is possible, they have been taking a holiday from the disease. Nic has been home for six weeks. In October, he turned five years old. He has been well enough to go to school, play, eat. It sometimes seems as if he is making up for lost time. He wakes hungry and asks for food all day long.

Amylynne doesn't mind. She and Sean have treasured these weeks with their illusion of normalcy. Although they try not to dwell on it, they know the dragon is only sleeping.

During the sequencing of Nic's DNA, Alan Mayer and the other doctors have been careful not to inflate the Volkers' hopes. They have reminded the family that in all likelihood this final test will produce the same answer as the others, no answer. Amylynne listens to their hedging and maintains her own quiet confidence. She is quite sure that DNA sequencing will deliver the answer to

Nic's mystery. She has even allowed herself to consider where the child's problem originated. Not with her genes. No, the problem had to have come from Sean's.

On a Friday afternoon in mid-November, Amylynne's cell phone rings.

It is Mayer. He says the doctors are excited. They have found the mutation—maybe. Ever since Howard Jacob agreed to sequence Nic, Mayer has felt confident. He will know the right mutation when he sees it. Something in the genetic script will stand out. His patient's problem will make sense.

Though he is now convinced they have found the cause of Nic's illness, Mayer is careful not to put it that way. He controls his voice, muting his own excitement. The Volkers have been through too much already. The last thing they need is false hope.

On the other end of the phone, Amylynne's mind is skimming over the new information. *The mutation . . . maybe . . .*

She wants to know everything, no holding back.

"All right," she says. "What disease? How many years does Nic have to live? Tell me the bad news first."

Mayer explains that a mutation on the X chromosome appears to have caused the illness in Nic's gut. But there's more: the same mutation is responsible for another illness that Nic has had all along, one the doctors did not discover until now. The new diagnosis is not official yet, and Amylynne seizes on that small degree of uncertainty, the chance that when they go to confirm the diagnosis, doctors will find something to make them change their minds. Much as she has longed for an answer, she now pours herself into the hope that this will not be the final answer.

If the new information is confirmed, it will mean that Nic has two rare diseases instead of one, and either could kill him before he reaches adolescence. The mother has much to process. Yet the repercussions of Nic's sequencing are only beginning.

The Volkers now face one of the dilemmas of the Genome Age:

they are about to learn that the information in our script is, in some sense, not ours alone. David Dimmock explained all this before Nic was sequenced, but it was overshadowed then by the family's desperate need to understand the child's disease. Nic's DNA carries implications for others in the family, and the sequence has now brought the Volkers and their doctors to a question Amylynne has only casually considered: Where did Nic's mutation come from?

Some mutations are inherited from our mother or father. Others are not inherited but caused by errors in the copying of genetic material or in cell division. Sorting out how Nic got his mutation is important because it may reveal whether his mother, father, or sisters share it.

The news about Nic's mutation could divide the Volkers, not only into those with and without the mutation but also into those wishing to know their status and those preferring to live without that knowledge.

To this point, the thorny questions of the genome have been largely theoretical, meaty topics for discussion at scientific conferences and in journals. Scientific bodies have yet to come up with an agreed-upon method for exploring these questions. But everything has changed with Nic Volker. The issues have vaulted from theory to reality. There is little time to pause and think; Nic's disease is forcing his family and his doctors to make choices in real time.

For that reason, on the morning of November 18, a few days after the call from Mayer, Amylynne meets with Dimmock to go over what scientists have found and the decisions that must be made. Dimmock tells her about the mutation, the two diseases (the second now confirmed), and the need for a bone marrow transplant, at least to treat the second disease. As for the first disease, the holes in the intestines, they must hope the transplant will cure that, too. But before the scientists can be certain about Nic's mutation, Dimmock says, they must send his blood for testing to a federally approved clinical lab.

Dimmock then tells Amylynne that they would like to test her blood as well to determine if she shares Nic's mutation. The geneticist proceeds carefully, explaining that what they find could raise questions for Amylynne, her children, and other relatives. Knowledge of Nic's mutation will loom over their family-planning decisions for generations to come. Dimmock presents Amylynne with the consent form that she must sign in order to have her XIAP gene sequenced. The questions on the form give a sense of the unusual path she is choosing in allowing science to peer into her genetic makeup. Dimmock runs through the questions out loud.

"Are you Nicholas's mother?"

Amylynne is surprised by the question. *Really?* she thinks, *You're asking me if I'm the mom?*

"Is Sean, Nicholas's father?"

The answers are: yes and yes. Of course they are.

But the need to ask such questions—odd and uncomfortable as they are—is not far-fetched. Somewhere around 3 to 4 percent of DNA tests end up inadvertently revealing nonpaternity. In other words, you start out looking for the Huntington's gene and learn that the person you've been calling Daddy, strictly speaking, isn't. DNA tells our secrets.

At the end of the consent forms, Amylynne must answer two final questions: having seen the implications of testing, does she still want her own DNA sequenced, and does she want to know the results?

Yes and yes.

For now, Amylynne has little time to fret about where the information from her own script may lead. Her immediate concern is Nic. The destruction and rebuilding of his immune system in September ushered in a period of remission. For a few months, he ate real food. But by early December, the remission appears to be over. Nic is losing weight again. The little boy, so full of energy after all those months cooped up in the hospital, is suddenly lethargic.

The news has been full of reports of swine flu, and Amylynne wonders if Nic might have caught it. Perhaps he has mononucleosis. That would be devastating news, as most cases are caused by the Epstein-Barr virus, the illness that kills boys with XLP.

Or maybe, as Mayer believes, allowing Nic to eat again has simply fed his disease, colonizing his intestine with fresh bacteria and sending the immune system back into a frenzy. If the doctor is right, the recipe for remission is a familiar one: intravenous feeding only, bypassing his gut. Mayer would then administer antibiotics to reduce bacteria in the intestine, and the drug tacrolimus to suppress the immune system. Unfortunately, it is not a long-term solution. Eventually the intravenous feeding will lead to chronic liver disease, and the tacrolimus will damage the kidneys.

Amylynne takes her son to Children's Hospital once again, hoping that whatever the cause of his current illness, it will not mean having to spend another Christmas there. She is not optimistic. She moves back into the Ronald McDonald House near the hospital. There, in her room, she buries her head in a pillow.

"This must be what hell is like," she writes in her journal.

Nic's ileum, the final section of his small intestine, is overrun with pus and ulcers. She worries he will never eat real food again. Then, a few days before Christmas, Nic erupts. "I want my food back! Give me my food back!" he screams. "I don't want to get better! I want to be sick and have my food!"

Amylynne has known all along how badly he misses food, but she has never heard him express it with such raw anger. She tries as much as possible to give Nic what he wants—a toy he sees in the gift shop, quality-of-life trips outside the hospital. The prohibition on real food is one rule she cannot bend. His cries for food are the wish she cannot deliver.

Amylynne knows there are worse things than another Christmas at Children's. In the last year she has grown close to other families with sick children, then watched some cope with the loss of their child.

"We could make Christmas in the Hospital bearable, even fun," she writes in her journal. "Heck, I can be happy almost anywhere, but on top of taking away Nic's home, his family and his activities for Christmas, the disease has taken away his food and his drink. No Christmas steak with A1, no Xmas cookies, no Chocolate Santas."

While Nic grapples with the latest onslaught from the disease, the scientists push hard to confirm that they have found the culprit. There remains a critical step in determining that the sequence on Nic's XIAP gene is to blame. Worthey and Dimmock have shown that the mutation is causing Nic's production of the XIAP protein to be just 60 percent of normal, but that does not prove it is causing his disease. Many people make less of a protein than normal and still live healthy lives. The scientists at Children's Hospital and the Medical College need to show that the mutation makes Nic's cells defective. They must conduct lab tests to demonstrate that the error in his sequence alters the function of the XIAP gene and produces the dysfunction in his body. They know what the gene is supposed to do in order for a person to be healthy, so they can observe Nic's cells in a lab dish and see whether the gene does its job.

Immunologist James Verbsky, the skeptic who once believed that sequencing would send Nic's team down a scientific rabbit hole, now finds himself in the position of designing tests to prove that the technique has worked. The evidence so far has impressed him. The news that a mutation has been found, a mutation so far unseen in nature, fills him with newfound excitement. They may be participating in the emergence of a powerful medical tool.

To this point, Verbsky and his fellow immunologists have run numerous tests focusing on single genes that appeared to be promising suspects. But each of these single-gene tests took as long as two or three months and cost as much as $3,000. Each time a test ruled out just one gene, immunologists had to begin the cumbersome process all over again. Sequencing all of the genes at once, as the scientists have done with Nic, essentially combines all those previous tests. Applied across the health care system, it could reduce the time and cost of a diagnosis and increase the likelihood of success.

In order to prove that sequencing has revealed the cause of Nic's disease, Verbsky designs two experiments that seek to determine whether the child's genetic "typo" damages the effectiveness of the XIAP protein in his body. When the XIAP protein is doing its job, it has two roles: blocking the process of cell death and preventing the immune system from attacking the intestine when food enters. The latter role is that of a gatekeeper, allowing good microbes to enter while destroying bad ones.

Verbsky stimulates some of Nic's cells in a lab dish by adding a bacterial product. If the cells are working properly, they should respond to the bacterial product as cells from any other human would: by releasing the protein IL-6, which is important in inflammation. Verbsky runs the test three times with Nic's cells and three times with other human cells. Each time, the other cells release the protein. Each time, Nic's cells do not.

In the second test, Verbsky and his colleagues try to determine whether Nic's XIAP protein is curbing the process of cell death. Once again, the boy's cells differ from other human cells. More of his cells are dying; the protein isn't saving them as it should.

Verbsky is convinced. They have now proven the case for XIAP.

Chapter 20

Faith and Doubt

JANUARY 2010

Three years of medical emergencies, puzzled doctors, and intravenous feeding lines have worn down the Volkers. At times Sean retreats into his own thoughts, silent and distant. Amylynne cannot reach him. For her part, Amylynne spends so much time at the hospital that she feels herself losing touch with life at home, where Sean is caring for their three girls. The family finds comfort in small things: a program that offers Nic a school of sorts inside the hospital; a day when the girls visit and play with their brother; news that Nic's ulcers have healed and he is gaining weight. By the beginning of 2010, the doctors' strategy of limiting his nutrition to intravenous feeding has succeeded in keeping the disease at bay.

"God is AWESOME!!!!!" Amylynne writes in her journal. "Where else could this have come from? Only from above."

In January, Amylynne returns to see David Dimmock. Nic's

mutation has been confirmed at a federally approved lab at Seattle Children's Hospital. The test in Seattle provided Dimmock with a graphic reminder of how far out from existing medical practice they have stepped in sequencing Nic. The lab official in Seattle is stunned to learn that a hospital in Wisconsin has sequenced all of a child's genes. It is a complete reversal of established procedure. Normally, doctors decide that a child appears to have a certain disease, then send out the one or two genes crucial to that disease for testing. They don't test all of the genes, then figure out the nature of the disease. Few doctors outside the Medical College of Wisconsin know about the sequencing of Nic Volker, and the reaction from Seattle is a taste of what will follow when the news reaches geneticists around the world.

But on this day, Dimmock's news concerns Amylynne's genetic script, not Nic's. Results have come back from the sequencing of her XIAP gene and from tests on her immune system. Dimmock and genetics counselor Regan Veith have carefully rehearsed what they will say. The doctor's role is to explain the diagnosis, Veith's to explain what it means. Veith has received special training for this work. At the University of Wisconsin–Madison, she ran through practice scenarios with trained actors on her way to earning a master's degree in medical genetics. Still, she is nervous as they walk down a second-floor hallway in one of the hospital buildings to a small exam room.

Veith knows there is an unavoidable tendency to look upon flaws in our DNA as flaws in ourselves, personal failings. The counselor tries to put herself into Amylynne's position. The mother's whole world is completely different now, entering the room, than it will be an hour from now, when they've finished talking and she walks out the door.

The exam room is bright and cheerful. Cartoon figures smile from the walls. A large window looks out on the hospital campus and the Ronald McDonald House, where Amylynne has spent so many weeks. Veith, Dimmock, and Amylynne sit together in a tri-angle, so close their knees are almost touching. Dimmock runs through a preamble, much the same as the one he gave a few weeks earlier when Amylynne agreed to have her XIAP gene sequenced. His words amount to a signpost for humanity as we enter a new age of genomic medicine. "We have no say in the genes we pass to our children," he explains. "There are benefits to reading our DNA, but they are balanced by the potential harm of what the DNA tells us."

The preamble is a big picture of the world from far above the little exam room, but Dimmock brings the conversation back to earth, back to Amylynne and what the doctors found in her ge-netic script.

"You carry the mutation," he says.

Like all females, she has two X chromosomes. One has the nor-mal XIAP gene. The other has the mutation. The good gene ap-pears to override the bad one. That's why Amylynne can eat food without developing the inflammation and holes in the intestine that plague her son.

Like all males, Nic has only one X chromosome; it is the one he inherited from his mother, the one that has the mutation. For him, there is no good version to override the bad.

Amylynne's eyes fill with tears. She did not get the disease; she just passed it to her son.

She understands that we do not choose our DNA, but that knowledge does not ease her mind. Even if the guilt she feels is not rational, it is real and it strikes at her image of herself as the warrior mom. She has been crusading to help Nic fight a disease that came from her. Veith slides over a box of tissues. Dimmock and Veith give the mother time to absorb the news in silence.

As a geneticist, Dimmock often must deliver devastating news to parents. Three or four times a week he is the messenger of heart-break. It is part of his job, and it has to be done carefully. "Research has shown that how parents bond with their children and cope with grief is so dependent on doing this right," Dimmock says.

He does not hug Amylynne, but his voice is gentle and reassuring. He and Veith steer the conversation toward the next stage in the process, the decision to treat Nic with a bone marrow transplant. That decision rests with the family alone.

Although this is the treatment for which Amylynne has prayed, there is nothing simple about the recommendation. The take-home message conveys a sense of risk and uncertainty: your son has something medicine has never seen before, and we are recommending a treatment that could kill him.

Officially, Nic's disease has no name. Doctors have taken to calling it XIAP deficiency. In a way, however, the second disease, XLP, is the more crucial diagnosis because it has opened the door to a treatment the doctors can justify.

Although the bone marrow transplant seems warranted now, Dimmock still feels a certain hesitation. He knows the transplant will treat the XLP—but what will it do for the first disease, the inflammation and holes in the intestine? He isn't sure. He isn't entirely certain that Nic's mutation is responsible for the inflammatory bowel disease. It's possible that it is only to blame for Nic's weak natural killer cells.

In January, Dimmock, Mayer, and Verbsky meet with David Margolis, the bone marrow transplant doctor. They explain what they have learned about the child's mutation and the second disease. They believe a bone marrow transplant is justified. The theories sound impressive, but before Margolis will perform the transplant, he must have a diagnosis in writing. "If you are certain he has XLP," he says, "one of you needs to put that in writing in the medical record."

Dimmock understands that they have reached a critical point. They are being asked to stand behind what they found through sequencing and their own analysis. For now, he sets his doubts aside. He writes a note affirming Nic's diagnosis of XLP and recommending a bone marrow transplant. Then he signs his name.

Although bone marrow transplants have come to seem almost routine, there are still only about eighteen thousand in the United States each year. Survival rates vary according to the disease the transplant is being used to treat. But the process carries an undeniable risk. Before injecting the cells that will create a new immune system, doctors must wipe out the old one, leaving the patient with no defense against infection. Also, the new, transplanted cells could attack the rest of the body, a potentially fatal condition called graft-versus-host disease. Nic's generally poor health makes the transplant even more risky. Dimmock and the other doctors know there is a very real possibility that in trying to cure him, they will kill him instead.

For their part, the Volkers are in no hurry to rush Nic into a transplant. Doctors at Children's recommend that the family seek a second opinion. The second opinion would be based on Nic's medical history and now on the information in his genes as well. Amylynne begins considering other institutions: Duke, the hospital in Cincinnati, hospitals in Minneapolis and Seattle. But there is a problem with going out of state. Since Nic has exceeded his $2 million in lifetime health benefits, he is on Wisconsin's Medicaid program. The program will determine where the Volkers can go to for the second opinion.

All of this means the bone marrow transplant must be put on hold, which suits the family just fine. While they determine where to go for the second opinion, Nic continues to improve, his weight finally reaching thirty-eight pounds—still only in the twentieth percentile for his age but a big step for him. He enjoys the school program run by Children's Hospital and the visits by his dad and sisters, when he can play monster trucks and jump on his bed.

The waiting period for the transplant sometimes brings Amy-lynne and the doctors into conflict. Knowing that her son may not live through the transplant, the mother tries to give him trips outside the hospital, quality-of-life days, she calls them. Mayer, however, is determined to control Nic's environment as much as possible, to keep him healthy until the transplant. One flash point is a trip the Make-A-Wish Foundation is providing Nic and his family.

Nic wants to swim with dolphins and go on a Disney cruise, but Mayer says that going into the water is not an option. The boy has a central line entering his body to deliver medicine and fluids. Patients with central lines cannot go swimming. Bacteria in the water can get into the line and cause disease. The doctor is firm: Nic has to come up with another wish.

When Amylynne pushes for Nic to get his wish, Mayer suggests that maybe she is thinking more about herself than about what is best for Nic. Maybe, he says, she is trying to give Nic these special moments because, privately, she has given up hope that he will survive. The very notion of Nic receiving a wish from the organization upsets Mayer. Wishes are granted for children who are dying, and the doctor is not ready to concede defeat.

"He told me that maybe I have prepared myself for Nic's death prematurely and that's why I want 'quality of life days' for him," Amylynne writes in her journal. "Honestly, can one ever prepare themselves for their child's death? Uh, I don't think so—but when one has a chronically sick child, I think it makes us want to live more, not less, and not sweat the small stuff."

After some debate, the doctor and mother reach a compromise: no swimming with dolphins, but a trip to Las Vegas and the monster truck championships instead.

The trip grows in importance for the Volkers as Nic's Christmas hospitalization stretches past two months. In the three and half years since his illness began, Nic has spent more than six hundred days in the hospital; he's been there as often as he's been at home. The time away from home drains Amylynne. She finds the process of getting Nic discharged complex and frustrating. Late in February, it appears he will be discharged, but some of his numbers have Mayer worried and the doctor orders a liver biopsy. He fears Nic's gut disease may be reawakening, but nothing of concern shows up on the biopsy. Now Mayer wants to examine the boy's ileostomy, the section of the small intestine that has been routed outside of his stomach to allow waste to pass into a bag. He is going to check each of Nic's organs one by one, Amylynne fears, then start over again. To the mother, it feels as if Mayer is trapped in an endless cycle of worry, his fear making a prisoner of Nic. As Mayer sees it, there is no point letting Nic go home if he is going to spend a few days there, then plunge into another crisis. So the doctor orders a procedure called an endoscopy, which involves threading a tiny camera and fiber optic light into the small intestine. Mayer makes no promises, but Amylynne cannot help feeling that there is an unstated bargain: if the doctor finds nothing wrong, Nic is free to go home and then on the Make-A-Wish trip to Las Vegas.

The endoscopy reveals no ulcers or holes. And it serves as the all-clear the Volkers have been praying for.

Nic enjoys the respite from disease. At the end of February, he returns home for the first time in more than two months. To Nic, being home means running around, bellowing "Freedom" at the top of his lungs. To Amylynne, being home means having an unaccustomed feeling of privacy—no doctors or nurses entering the room whenever they please.

In late March, as they are packing for the Las Vegas trip, Nic's lab tests show his inflammation levels rising sharply. The disease is entering the child's ileum, the final section of the small intestine. Mayer knows how important the trip is to the family. He knows, too, that the presence of the disease in Nic's ileum is serious. If it is destroyed, the boy will not make it to adulthood. There are no ileum transplants.

On a Sunday, while Amylynne is in church, the doctor emails asking that she contact Medicaid to get advance approval for lab tests at a hospital in Las Vegas. That way, Mayer can check Nic's condition during the trip. The day of the trip, even as they fly toward Vegas, Amylynne is worrying, taking Nic's temperature. He has a low-grade fever. But it would take a full-blown emergency to cancel the trip at this point.

Las Vegas does not disappoint. The skies are cloudless. Nic holds a baby lion. He gets a visit from Batman in the hotel room. His eyes and ears fill with hours of monster truck madness— a world apart from the hospital, with its buzzing and beeping machines. His disease, mercifully, does not flare up.

After a few days of vacation from the hospital, they return to Wisconsin and begin the final preparations for the bone marrow transplant. A second opinion from the University of Minnesota supports the decision Sean and Amylynne have already made. Although they fear the transplant, it seems Nic's best and only option.

Among the doctors, there has been a sense of urgency. Every day that the family waits is another day that Nic could be exposed to the Epstein-Barr virus, one of the most common viruses on earth—an inconvenience for most, but fatal for Nic.

The three years of Nic's illness have intensified Amylynne's faith but also made her sensitive to what she calls "signs and wonders." She pays close attention to the language of Sunday sermons, the dreams that pass through her subconscious, even the animals and scenery she sees as she jogs. She is hyperaware of anything that might be a portent about Nic and the transplant.

In June, a few days before the transplant, she dreams of the word she has been so careful not to speak or write, the "D" word. In the dream, Nic lies on the ground. His blue eyes are closed, and there is a peaceful smile on his face. The heat is stifling. A boy and girl stand over Nic, pointing down and holding their noses. Amy-lynne crouches beside her son, gripping his shoulders, crying his name over and over. He does not stir.

"I can still hear myself screaming," she writes in her journal. "I hope and pray this is just my mind playing tricks on me."

Chapter 21

The Smell of Creamed Corn

JUNE 2010

It is Sunday, the day after Amylynne's nightmare. It is also Father's Day. Tomorrow, Nic is scheduled for surgery. Doctors plan to install a central line that will allow chemotherapy to seep into his body, destroying his old immune system and preparing him for the bone marrow transplant. The Volkers have one day to ready themselves for the start of the grueling, risky procedure.

Now, though, for the first time in four years, Nic is home with his family on Father's Day. The Volkers drive to Damascus Road Church, which sits in a quiet strip mall in Madison. Sean carries Nic inside, slumped over his shoulder. They want so badly to spend the day outside the hospital, but it isn't looking good. Dressed in his favorite Batman cape, Nic keeps closing his eyes. His face is ashen; his body limp.

Friends greet the family and ask about Nic, who has been unable to attend church for many weeks. As the service begins,

Amylynne stands, her eyes closed, arms outstretched. Sean remains seated, Nic's head buried in his chest. Daniel Peck, a friend of the Volkers, requests that the congregation pray for the family. Then the Volkers go to the front, and Amylynne begins describing Nic's struggles and the ordeal he is facing.

"Pray against complications," she asks as Nic rests against his father's shoulder. "Pray that the new immune system will take, because Nic—"

Her voice falters and she cannot finish. Members of the congregation press forward, surrounding the family, placing their hands over Nic's head. A red spotlight bathes the child and the arms reaching toward him.

"We have your favor and we know that, God," prays Joseph Steinke, a cofounder of this evangelical church. "But we live in this broken and disrupted, unhealthy world . . . Jesus, we're asking you to come and to place your hand on little Nic, and would you just guide him through this next week."

When the prayer finishes, the Volkers depart. Over the next hour, Nic's fever passes. They don't have to spend Father's Day at Children's Hospital. They get to play minigolf instead. That night, though, Nic has trouble sleeping.

"Put your arm around me," he tells his mother.

The next morning, Amylynne does not take Nic to the waiting area for his surgery; she carries him straight to the emergency room. The color has drained from his face. His fever is back.

"Water, Mommy. Get some," he pleads. "Get some. Get some."

But he cannot have water. There is still a possibility he will have surgery, and if he does, he cannot have water beforehand.

The decision about the surgery rests with David Margolis, the doctor leading Nic's transplant. Margolis is a great believer in the fundamental principle of medicine: Do no harm. He is also a stickler for proof, a trait instilled in him in childhood by his father, a lawyer. It was Margolis who refused to approve Nic's bone marrow transplant until the other doctors proved to his satisfaction that they'd found the cause of the boy's illness and were confident enough to sign the diagnosis.

Margolis examines Nic and postpones the surgery. The child has sepsis, the same hyperimmune response that almost killed him in 2007. By nightfall he is resting in intensive care, his mother hovering at his bedside. She has seen him this gravely ill before.

In the morning, Nic sits up in bed and kisses her.

"Mommy," he says. "I had a bad dream last night. A scary ghost came into my room to get me."

Nic explains that in the dream there was also a young man in the hospital room who promised to protect him.

Amylynne's faith shapes her interpretation of the dream. She believes the young man is Jesus and his appearance a sign that Nic will pull through. To another mother, such a dream might seem a fragile thread on which to hang hope, but Amylynne is happy to have it. She feels comforted.

In a few days the sepsis clears, and the countdown to transplant begins once more. The procedure is scheduled for July 14. Margolis has decided that Nic will receive umbilical cord blood from an anonymous donor. Cord blood is similar to bone marrow, but Margolis prefers it for Nic because it carries less risk of graft-versus-host disease, the condition in which the new cells attack the rest of the body. Even so, the doctor is careful to explain that cord blood only reduces the risk. Many things can go wrong.

The Volkers do not dwell on the dangers, especially when they talk with Nic. The idea of the transplant is difficult enough to explain to a child. Adults who receive transplants of cord blood or bone marrow often come to think of the event as a second birthday, the day their blood is reborn. Amylynne explains this idea to Nic, but it turns out the child has fashioned his own way of looking at the transplant. Until now he has been Batman. He has worn his cape and mask and gloves into the operating room. He has even insisted that doctors and nurses call him Batman instead of Nic. Recently, though, during a brief trip outside the hospital, he saw a sneak preview of the movie *The Last Airbender*. Nic was very taken with the character of Aang, a boy with a shaved head who must battle powerful forces. Nic explains to his mother that he will go into the transplant as Batman and emerge as Aang.

The transformation is already under way. A week of heavy-dose chemotherapy leaves his head bald like Aang's. Though the chemo makes him vomit, he proves remarkably tough. "Mom, I just got sick and I took care of it myself," he tells Amylynne one day when she returns to the hospital room. "I got a bucket myself."

The afternoon of the transplant, July 14, begins poorly. One of the nurses has made a special Batman poster for Nic, but the gesture fails to soothe him.

"Go away!" he screams before the transplant has even begun. "I don't like the Batman sign. Take it apart!"

Amylynne gently reminds him of the transformation he is undergoing—from Batman to Aang. But Nic howls. He will not be consoled.

At 2:17, with a storm brewing outside, the procedure begins. Crucial as it is to Nic's survival, the transplant seems surprisingly mundane. In early 2009, Nic went through dozens of wound cleanings, each one lasting two hours, more than twice as long as the transplant that could now save his life. There is no anesthesia, no elaborate preparation, just a small bag of cord blood from an unknown donor. A nurse starts the flow, and the rose red liquid seeps from a 50-milliliter syringe through a clear intravenous line into a vein in Nic's chest.

"We're starting, Bud," Amylynne says.

Although Sean cannot leave work to come to Children's Hospital, two of Nic's sisters are in the room. Amylynne lies in bed beside her agitated son, just as she has so many times over the last few years. As she holds him, she thinks how odd it is, the smell from the cord blood. It reminds her of creamed corn.

Nic slowly settles down, his cries fading to a whimper. He reaches over and pulls his mother's arm across his chest. His face calms. His breath becomes slow and rhythmic. Amylynne has prepared for this day, right down to the sound track she has brought with her. She begins with "Spring" from Vivaldi's *Four Seasons*. Violins fills the room, and the mother stays beside her five-year-old son, watching as he drifts off to sleep.

Twenty-eight minutes into the transplant, Amylynne begins to weep. "I pray," she says, "that sickness and infirmity will have no place in his life." She moves on to Psalm 107, beginning at verse 20: "He sent forth his word and healed them."

At thirty-five minutes, Vivaldi is replaced by "Aang's Theme," from *The Last Airbender*.

At forty-eight minutes, the last of the cord blood enters Nic. It is just past three in the afternoon. After a few minutes, he sits up.

"How do you feel?" his mother asks. "Any different?"

"Nope," he says.

The transplant is only the beginning. The days that follow are tense and filled with worry as doctors wait to see how Nic's body handles the new immune system. So great is the uncertainty surrounding the transplant that doctors cannot agree how many days must pass before they can feel confident that the procedure has been a success and Nic is home free. There are lengthy discussions about what the right time span should be: fifty days? a hundred days? longer?

Later, Dimmock will recall that the informal consensus settled on a hundred days. In a real sense, one hundred days is a fairly arbitrary number, based mostly on the fact that many of the complications from a transplant can be traced to the high doses of chemotherapy used to knock out the old immune system. The worst effects of the chemotherapy tend to show themselves within those first hundred days.

Following the transplant, Nic is energetic and moody. He wants to play. He does not realize how vulnerable he is. Eight days after the transplant, his old white blood cells, the body's defenders against infection and foreign material, are gone. The new cells from the transplant are still journeying into the hollows of his bones, where they will begin making red and white blood cells and platelets. For now his body is without a working immune system. It is virtually defenseless. Sores form on his mouth, throat, tongue, gut; they are the result of mucositis, a common reaction to chemotherapy. Nic loses his patience.

"Stop talking!" he barks at his mother. "Don't ask any more questions."

That night torrential rains whip through the area, and Nic's fever rises. His heart rate and blood pressure climb.

"Something is brewing in his body," Amylynne writes in her journal.

Doctors aren't certain what they are battling this time. There is talk that Nic may have sepsis again. Maybe a fungal infection. It is almost a week before doctors name Nic's new foe: adenovirus.

The virus strikes his respiratory tract. In a healthy person, the symptoms might seem like those of an ordinary cold. But in a patient with a weakened immune system, the virus can kill. Despite his deteriorating condition, Nic is full of fight. His oxygen levels have been dropping, but when the medical staff attach an oxygen mask, he screams that he is fine. Then he pulls out the cord.

Five days later, doctors are investigating a longer list of possible complications: sepsis, fungal infection, graft-versus-host disease, engraftment syndrome, human herpesvirus 6 (HHV-6). They fight back by treating Nic with a variety of powerful antibiotics, large doses of IV steroids, and almost daily transfusions of blood and platelets. The boy's moods swing sharply, a phenomenon Amylynne dubs "Roid Rage." He continues to insist he is not sick. He demands toys, and his mother is in no mood to deny him.

In her journal, she writes, "What do you say to a kid who has lived over 600 days in the hospital, has lived without food more of his life than with it, who has had over 150 trips to (the operating room), who has just had a transplant and can't go out of his room—now not even at night when everyone else is sleeping, even with a mask on?"

Each time, the doctors suggest a new complication, Amylynne busies herself searching for information on the Internet. Just as she hates to be surprised by news about Nic, she also hates to hear the doctors discuss a condition unfamiliar to her.

The Volkers and their medical team count the days after the transplant. By day 20, the adenovirus is advancing. But there is good news too: Nic's white blood cell count has risen. The donor cells are engrafting. The new immune system is taking root.

Day 24: The adenovirus is in retreat. The white blood cell numbers keep growing. A pudgy smile lights up Nic's face.

Day 29: Amylynne takes to her online journal and issues an urgent request for prayers. Nic has a fever. He is struggling to remember things, even his mother's name. He tests positive for HHV-6. This virus can lead to a host of new threats, including encephalitis, an inflammation of the brain. Mention of the brain fills Amylynne with dread. She has visions of Nic saved but profoundly brain damaged. His deteriorating memory only reinforces her fear. No sooner does Nic hang up the phone in his hospital room than he forgets which of his sisters was on the line.

Day 31: Tests confirm that Nic has encephalitis. He lies in bed, his face blank. This was not supposed to happen. HHV-6, encephalitis, memory loss—none is a normal complication of a cord blood transplant.

"There is no fight anymore with his cares," Amylynne writes, referring to the routine tasks the nurses perform, such as taking blood and checking vital signs. "He just lays there, all docile and good. I absolutely hate it and it breaks my heart."

She misses the Nic who growled when doctors and nurses annoyed him. She has difficulty sleeping. She worries that whatever is happening in the brain will rob her son of his personality. One day she says, "If I knew this was coming, I probably wouldn't have done [the transplant]. That's how I feel today."

The doctors cannot focus on the brain because Nic's body comes under a continual onslaught. He gets a rash. He gets graft-versus-host disease. He gets a staph infection.

The medical team hustles to anticipate each new complication and respond. The doctors give Nic antivirals for the adenovirus, HHV-6, and encephalitis. They prescribe antibiotics for the staph infection. They combat graft-versus-host disease with immune-suppressing drugs.

After two anxious weeks, Amylynne sees a hopeful sign: Nic snaps at the anesthesiologist for daring to touch one of his stuffed animals, a bull.

Day 47, August 30: Nic's memory is improving, and he is well enough to eat real food for the first time in months, chicken noodle soup. The simple pleasure of a bowl of soup prompts the boy to declare this day the happiest of his life. Two weeks later comes an even better moment. Nic gets to eat the food he has told nurses and doctors that he craves most of all: steak.

Amylynne feels a happiness she has not experienced in a long time. Her words on Facebook soar gloriously over the top. "The heavens are opening up," she writes, "the angels are singing and Almighty God is dropping down bottles of A.1."

There are still a few bad days in September—an infection from an intravenous line and migraine headaches—but Nic keeps improving. Tests no longer show any trace of the adenovirus or HHV-6. His white blood cell count rises and is just a shade below normal by the middle of the month. The new immune system has established itself, and the Volkers are able to begin preparing for Nic to leave the hospital.

"He finally got his discharge," Amylynne says, choking up.

Day 99: The big day arrives at last on October 21. There is so much to do, the packing, the instructions for Nic's discharge and the good-byes. It has been 120 days since he entered the hospital. Nic has spent more than half of the year at Children's Hospital, but he is going home in time to celebrate his sixth birthday, October 26. His room at the hospital looks like any other boy's bedroom, an accumulation of toy monster trucks, clothes, games, and Batman gear that Amylynne packs in plastic tubs, then hauls out to the family minivan.

David Margolis, the transplant doctor, enters the hospital room and gives Nic the final OK to leave, then a high five.

"Nic," the doctor says. "Thanks for coming. Thanks for leaving."

Still vulnerable to infection, Nic wears a mask as he walks through the halls, headed for the exit. Just before he and Amylynne leave for a home life they've almost forgotten, Margolis comes rushing out of the hospital. He is accompanied by a stranger in a shirt and tie. It has been more than a year since an email from Alan Mayer proposed sequencing the child's DNA, and in all that time the man who gave the critical go-ahead for the groundbreaking procedure has never met the boy.

Howard Jacob has thought of Nic many nights and prayed for him. Nic has been the subject of talk at the scientist's dinner table. He has come to know a great deal about the child's genes, the code tucked inside each one of his cells; yet he and the boy are strangers. This is the first time the scientist has seen his patient in the flesh.

"How are you?" he asks, leaning down toward the child.

Nic, armed with his plastic *Last Airbender* staff, asks in his high-pitched voice: "Are you tough?"

The scientist, unprepared for this line of inquiry, begins to explain that he is not so tough, not really.

Nic isn't waiting for the answer. After months cooped up in

the hospital, after all the days when the doctors and nurses jabbed him with needles and took blood and poked him whenever they pleased, he is free, his body brimming with boyish energy waiting to be unleashed.

Nic grips his *Airbender* staff and jabs the world-famous geneticist in the gut.

Chapter 22

Inheritance

2010–2014

N ic wastes no time diving back into his childhood. He lives now as if trying to compensate for all he missed in his years at the hospital. He thunders around the Monona house. He savors the pleasures of McDonald's Happy Meals. He enjoys *real* play dates—not the kind that take place in a hospital room. The winter of 2011 brings plenty of snow to Wisconsin, and he revels in the chance to do more than just stare at it from a window.

Prior to his release, Nic's doctors engaged in long debates over how much time it would take before they'd be able to say that Nic's mysterious intestinal disease had been cured. Though they developed an uneasy consensus around a hundred days, some doctors—Mayer in particular—remain hesitant to sound a medical "all clear." Nic's disease has fooled them before, and they don't want to be caught off guard.

Mayer's own benchmark is more conservative; the disease had

first struck two years into Nic's life, and Mayer thought of the cord blood transplant as a second birth. They would need to wait two years—more than six times the general consensus—before Mayer could be sure the disease wasn't simply lying dormant.

But the boy races through every checkpoint. He reaches two hundred days, one year, even Mayer's two-year mark—and still no symptoms of his initial disease reemerge. His gastrointestinal health is good, Mayer reports. No more painful holes in the intestine. He's able to eat whatever he wants.

In July 2015, Nic marks the fifth anniversary of his transplant. The disease has not returned.

Those five years bring much progress. In April 2011, nine months after his transplant, Nic is able to return to kindergarten. In many ways, it's the first real schooling of his life. Though he was enrolled in an early-childhood program in the early stages of his disease, when he was three, he was seldom well enough to attend. He had the same problem in four-year-old kindergarten, which he attended rarely. He wore a special backpack to hold the liquid nutrition that seeped through his intravenous feeding line. He never got to know the other children. All those years he was really enrolled in the hospital.

Now he resumes school. He gets invitations to other children's birthday parties for the first time. It has been years since he's been well enough to have a birthday party of his own. His life is just beginning to widen beyond the hospital into something like a recognizable childhood. He plays T-ball in the spring and summer and tries tennis, swimming, and skateboarding. Sean and Amylynne get choked up watching him. In previous years, even simple things such as going to the grocery store were a hazard; shopping trips often took place later at night, when there would be few people around and less risk of encountering viruses that might overwhelm Nic's immune system. Now trips to the store, to church, to class are the everyday events they should be.

Nic's progress has not been without difficulties, however. In the years following his transplant, his health remains a worry. He fights off another bout of sepsis, requiring almost a month in the hospital, and endures several shorter hospital stays. Beyond the medical complications, doctors suspected there would be other lingering repercussions from Nic's lengthy illness. Any long hospital stay tends to leave collateral damage. Nic remains smaller and grows more slowly than many of his classmates, a result of the long stretches of intravenous feeding. After missing so much school, he lags behind his peers in classwork and behavior and has few close friends.

"He never got the chance to be that little kid," Amylynne says.

Due to the colectomy, Nic will always require a bag to collect his waste. Though he's used to the bag, he's still bothered by it. No one else in his class has one. He sometimes wishes that the bag would disappear and the opening in his abdomen, the stoma, would close up, leaving his tummy as smooth as it was before the illness.

There is a less obvious difference between Nic and his classmates. As a result of the transplant, he has the DNA of two distinct people. Were Nic ever to commit a crime, he could leave two sets of DNA at the scene: his own, in a strand of hair; and that of the donor, in his blood.

The most alarming legacy of Nic's transplant arises a little over a year after he leaves the hospital. He starts experiencing seizures, often several times a day, and is diagnosed with Doose syndrome. An epilepsy disorder of early childhood, Doose is a rare side effect of the cord blood transplant. It is harrowing to endure, both for Nic and for Amylynne, especially after all they've fought through. He starts taking a medication called felbamate, which helps somewhat.

Seeking a better solution, Amylynne locks on to the idea that cannabidiol, or CBD oil, might help Nic more. She works with mothers of other epileptic children to get a law passed in Wisconsin that allows them to access the marijuana derivative, which doesn't provide a "high," but is believed to reduce seizures.

It remains unclear whether Nic will grow out of the seizures or whether they will be something he lives with into his adult years.

Amylynne sees that in many ways her son's life will continue to be shaped by the four years of his illness. In the fall of 2014, he is diagnosed with post-traumatic stress disorder. Some might view the diagnosis as a setback; but Amylynne greets it with relief. A diagnosis presents an explanation, which can, in turn, present an opportunity for healing.

After Nic completes eight weeks of talk therapy, Amylynne rejoices in a number of firsts. Her third grader sits through an entire movie without acting up for the first time since his transplant; he starts and finishes a school project with excellent comprehension; and he makes up with the child in day care he had viewed as his "number one enemy." Within a few months, he is able to walk with his therapist through the surgery clinic at Children's Hospital without tantrums or seizures.

Nic continues to improve. He enjoys class more, and he revels in the chance to play kickball with the other children. "Every day and every minute is a gift," Amylynne says. His appetite improves and his diet expands. His favorite food changes constantly—there's a potato chip phase, a beef jerky phase. He discovers guacamole, or as he calls it, "the green stuff that goes with chips."

The young boy lives for weekends, which bring rides on the family's fourteen-foot jet boat. Nic zips across Lake Monona, smile wide, lit up by the thrill of speed.

After the four grueling years of Nic's medical odyssey, the family too is healing. Marked by financial stress and long separations, the years spent living in the hospital have strained relationships close to the breaking point—something that's all too common among families facing an extended illness. By 2013, the cost of Nic's medical care reaches $6 million, the point at which Amylynne simply stops counting. She and Sean just keep making payments on the old bills.

The couple grew apart during Nic's illness, with Amylynne spending most of her time at the hospital and Sean spending his at home. "I had basically lived a year without him, and he'd lived without me," Amylynne says. "When I came home, I was ready to get a divorce. I think we were both ready to divorce. Then Sean got laid off. We basically had to spend time together."

Sean and Amylynne made a concerted effort to bridge the distance between them after Nic returned home. They had to relearn the dynamics of living together, getting Nic to school, shopping for food, cooking, and cleaning the house.

"We learned that, yeah, we really do like each other," Amylynne says.

The time that Amylynne spent away from Sean during Nic's illness, she also spent away from her three girls. She was missing in action for much of the girls' critical teen years.

The girls were shielded from many of the worst moments in the hospital when their brother hovered close to death. As Amylynne explains, "I didn't want them to become the siblings who lost their sibling. That was the pain I was going to spare them." But Amylynne's efforts, she believes, "probably backfired." Instead of feeling protected, the girls felt ignored. The reason for their mother's absence could never mean as much to them as the absence itself.

In the years immediately following Nic's crisis, the girls all struggle academically and socially. Leilani quits the pom-pom team. During her sophomore year of high school, she gains thirty pounds and endures bullying. She begins drinking and staying out partying.

In 2011, Mariah moves out. A year later, Leilani and Kristen do the same. As Leilani says, "By the time we realized how much of a strain it was on the family, it was way too late."

Still, they all work to ease that strain. Amylynne sends Leilani two dozen roses at school on her birthday, to her daughter's delight. And in 2014, the family rallies to celebrate Leilani's high school graduation. Now eighteen, she talks of becoming a pediatric surgeon, an idea she arrived at when her brother got sick.

Amylynne, a prolific Facebook writer, posts photos from the graduation, just as she did a few years ago with photos from the hospital. The intensity with which she joins Leilani's drive to graduate, an achievement not assured until the final week of school, is reminiscent of the determination Amylynne showed throughout the course of Nic's illness.

The Volkers have accepted, more than embraced, their roles as pioneers in clinical sequencing. Sean and Amylynne both have had all of their genes sequenced, largely to help doctors understand Nic's condition and where it came from. But no other family members have had their genetic scripts read. They have no desire to know what lies inside their genes. The Volkers remain convinced that the initial sequencing of Nic was the right decision. He was dying, and doctors could do nothing until they solved the mystery inside his body. Their only hope was that Nic's genes might harbor the answer.

Peering into the genetic script has proved to be a more troubling exercise for his parents. Amylynne already knows more about her own inheritance than she ever desired to know. Not only does she carry the mutation that made Nic so sick, she has risk factors for other conditions.

Sean is aware that his script contains warnings of its own, though he neither knows, nor wants to know, the specifics. He asked Amylynne to check his results. She waited at least a month before she told him what the doctors had found and even then explained only the general situation. He asked her not to provide details and she has honored his request, sharing the information with no one else.

"It's hard for me," she says, "because it's something that really should be shared with the entire family. But I promised Sean."

Though she hesitates to say it, the message she reads in both of their scripts is that they were never meant to have children. She has mentioned this conclusion to Sean, but they very seldom discuss it. If they talk about the more troubling aspects of their genomes at all, it is only in a light, half-joking way. So far, they have chosen not to talk about the subject in any depth.

"If I were to do it all over again," Amylynne says, "I wouldn't get my own information. It's disturbing."

Leilani agrees. She has told her mother she does not want to learn what's in her genome. When she begins dating a man seriously, she is terrified by the thought of having children and possibly giving birth to a boy. The odds are 50 percent that she carries the same mutation her mother passed to Nic.

"I don't think I'd want to put my family through what we've been through with Nic," Leilani says.

For Nic, it may be many years before he develops strong feelings toward his DNA. He had no choice in the opening of that dark box. Doctors continue to debate whether parents should tell young children what is in their genes. It is one thing for a parent to know and make medical choices to save a child's life. But suppose the script shows that a child is destined to have an incurable disease by middle age—should it be the child's decision whether or not to live under such a shadow? Should they be told by the time they are old enough to have children of their own? The genetic script continues to raise questions that defy easy answers.

Nic is unaware of the debate his treatment brought to the forefront of medicine. He just knows that what happened to him in the hospital made him famous and different from other children. "I'm a celebrity now. People know me," he says. It is half statement, half question.

"Kinda," Amylynne answers. "In the medical world."

Nic remembers a surprising amount about the hospital—the time he was dying from sepsis, the toys everyone brought, the names of the friends who visited and the doctors who treated him. He names his Japanese Chin dog Howard and sometimes calls him by his full name: Howard Jacob.

There are limits to what the genetic script can tell us, and there are questions that won't be answered for Nic until years from now. Children who go through bone marrow or cord blood transplants and the heavy doses of chemotherapy that accompany the process often have hormonal problems and sterility. They are at greater risk of cancer later in life. Nic faces all of those possibilities.

Later in life.

Amylynne still remembers the terrible nights when Nic lay near death in the hospital. "Later in life" was the answer to her prayers back then. Only now is it something she can take for granted, as other parents do when they think of their children growing up.

On October 26, 2014, Nic turns ten. For the first time in his young life, his parents organize a birthday party and invite his friends. Too often he has spent his birthdays in the hospital or sick or forced to avoid other children to protect his battered immune system. An annual ritual for other parents and other children, the birthday party is momentous for Nic. For weeks leading up to the day, he keeps saying "I just can't wait. I can't wait. I can't wait."

The party takes place at Amylynne's health club. Nic and nine other kids play kickball and dodgeball. They eat marble cake with vanilla frosting and even go swimming. During his illness, Nic could not go swimming, but he is well enough now. Some of the children come home with Nic and continue playing after the party. It is as near to a perfect day as the Volkers have ever had.

"It was huge," says Amylynne. "It brought tears to my eyes."

Nic loves to sing these days. The song "Let It Go" from the movie *Frozen* is one of his favorites. There is also a country music tune he has been partial to for a few years now, the Zac Brown song "Toes." Some days he walks through the house singing with a smile, one that widens knowingly when he reaches the inappropriate bits.

> *I got my toes in the water, ass in the sand,*
> *Not a worry in the world, a cold beer in my hand.*

Amylynne smiles, too. She likes hearing him sing the next line: "Life is good today."

Her family nearly fell apart, but it's now rebuilding. Leilani has moved back in, and the other girls are in touch. At Thanksgiving dinner, Leilani takes a video of her kid brother as they sit down to eat. For family celebrations, the Volkers always open a bottle of champagne. This time Nic pops the cork.

"It's Nic's story," Amylynne says. "But it's really our story."

Chapter 23

Come Along and Follow Me

2010–2015

Nic Volker's sequencing signaled the start of a new era in genetics, just as Howard Jacob had suspected it would. The ability to read our genes had been translated into medicine, revealing the secret of Nic's disease when nothing else could. Months after Nic's discharge in late 2010, Jacob launched a full-scale sequencing program, the first of its kind in the United States. Under the program, doctors at Children's Hospital began sequencing many more patients with mysterious diseases that had defied diagnosis.

Jacob and his team devised a set of criteria that cases would have to meet in order to qualify for this expensive test. They developed a protocol that included hours of genetic counseling. They pushed to get insurance companies to cover sequencing, an important step in helping the technology become accepted as the standard of care for treatment of unknown diseases. In his office, with its view of the emergency helicopters landing, Jacob left a simple message scrawled on the whiteboard to guide his team. It was a question: How many Nics?

After years of imagining the possibilities for genomic medicine, the hypothetical had been made real.

Mary-Claire King, the pioneering geneticist at the University of Washington who proved the existence of the breast cancer gene BRCA1, was struck by how rapidly the single accomplishment gathered into a wave. King, a friend of Jacob's, had played a major role in translating the genetic script into improvements in human health and a new understanding of our place in the world. Her doctoral work in 1973 showed that chimpanzees and humans are 99 percent genetically identical. When she learned what her friend's team had done, using a child's sequence to unravel the mystery of his disease, she knew it would be a turning point. She was impressed by the elegance of the work that had led to a diagnosis for Nic Volker. This was not the typical research project, examining a rare, genetic illness running through generations of a family.

"It was as if one had heard about Case Zero of AIDS and the cure, all at once," King would later say.

She had learned of the Wisconsin case a little before the rest of the world, at a genetics conference in late 2010 in Washington, DC. Six months later, she watched genome and exome sequencing burst into the forefront at a European genetics conference. It was not just one case now, King said, "but story after story after story."

As she left the conference, she observed to herself: "The world has changed."

Few would have bet on Jacob's lab to be in the vanguard. Jacob had not contributed to the landmark 2001 paper that presented a first draft of the sequence of the human genome. He had not been involved when Tom Albert designed the chip that would isolate Nic Volker's exons. He was not there when Jonathan Rothberg, the founder of 454 Life Sciences, conceived the then-novel idea of using computer chips to read the genetic script. Mary-Claire King herself would admit that when she heard the news of Jacob's accomplishment, her first thought was one of surprise. Over the years she had come to know Jacob as one of the world's authorities on the rat genome. *Howard*, she found herself thinking, *what happened to the rats?*

Yet there he was at the forefront of a new medical movement, among the first to demonstrate the power of sequencing to decipher an unknown disease.

Jacob's rat work had given him the expertise, but it was a more unique quality that propelled him into his pioneering role.

"Anybody involved in the Human Genome Project could have sequenced a kid faster," said Richard Roman, Jacob's former colleague who is now the chairman of the Department of Pharmacology & Toxicology at the University of Mississippi Medical Center. "But would you have found the risk taker who would have sent the resources to it? That is Howard's personality. He said yes."

As with many human milestones—the four-minute mile, the ascent of Mount Everest, the discovery of DNA's structure—multiple groups had the goal in their sights and were in hot pursuit. Doctors had used exome sequencing to make a diagnosis once prior to Nic Volker, though their work was not published until after the Wisconsin scientists had begun analyzing Nic's script. The case, involving a Turkish baby, was reported by researchers at Yale University in *Proceedings of the National Academy of Sciences* in November 2009—about the time Jacob's team was pinning down Nic's disease to that single error on a single gene. The baby had been suspected of having a disease called Bartter syndrome, a rare condition that can cause kidney failure. Turkish doctors had contacted Richard Lifton, the chairman of the Department of Genetics at Yale, because of his expertise in genetic diseases involving the kidneys. Lifton had been closely involved in the development of exome sequencing, and he saw the Turkish case as an opportunity to apply the new method.

"We were looking for a good opportunity to take the technology we'd developed for a test drive," he recalled.

Sequencing and analysis by Lifton's lab revealed that the Turkish doctors' diagnosis was incorrect; the baby actually had congenital chloride diarrhea, a serious but treatable disorder that can threaten an infant's ability to thrive. Unless adequately treated, most babies with the rare disorder die within a few months of birth. Fortunately for the family in Turkey, sequencing identified a disease the doctors had missed, allowing them to save the baby. According to Lifton, the child has survived. "We were convinced at that point that we ought to be using this technology as a discovery tool," Lifton said.

Despite Lifton's precedent, it was Nic Volker, not the Turkish baby, who would become the medical milestone. There are a number of reasons, including a seemingly small technical difference between the two cases. Yale treated work on the Turkish child as research, requiring that the project receive approval from the university's institutional review board, the group that governs research involving human subjects. That meant the sequencing was first and foremost an experiment, a chance, as Lifton put it, to take the technology for a test drive. In contrast, even though Jacob had been forced to tread delicately, presenting the work as both medicine and research, Nic Volker's treatment was always about saving the child's life.

Although other hospitals and institutions had hoped to use sequencing technology in the treatment of a patient, none had ever turned to it in such a life-threatening emergency. The publicity surrounding Nic's treatment further cemented his position as the face of a new kind of medicine.

Perhaps no one put it better than Eric Topol, the director of the Scripps Translational Science Institute in La Jolla, California. "Nic Volker," he said, "was a roar heard around the world."

Liz Worthey, the Scottish data expert on Jacob's team, took the lead in writing a paper for publication, the crucial step in certifying medical and scientific advances. It was accepted into *Genetics in Medicine* and published online in December 2010, about two months after Nic left the hospital.

The public news of Nic's sequencing, first in the scientific journal, then in a three-day series in the *Milwaukee Journal Sentinel*, became a watershed moment for genomics and personalized medicine.

Within weeks, the Volker story was everywhere.

In late December, Nic, Sean, and Amylynne were interviewed by Matt Lauer on *The Today Show*. Nic squirmed through much of it, and Sean left all of the talking to Amylynne.

Jacob's lab and the Children's Hospital doctors began receiving calls from desperate parents, asking if exome sequencing could be done on their children.

In February, Francis Collins, the director of the National Institutes of Health and one of the leaders of the Human Genome Project, lauded Nic's sequencing in an essay in *Science* titled "Faces in the Genome." Nic, Collins suggested, was flesh-and-blood proof of the genome's value. "My hope," he concluded, "is that when the day arrives to celebrate the 20th anniversary of the original human genome publications, we will be able to look at a world filled with the faces of people whose health has been improved by the sequencing of their genomes."

A few months later, Collins highlighted the Volker case in budget testimony before a Senate subcommittee. The National Institutes of Health announced that it was committing $416 million toward the goal of making analysis of the genetic script part of standard medical practice. Other institutions began sequencing programs of their own.

At a conference in Florida, Mike Tschannen's roommate discovered he was from the Medical College and asked if he had been involved in the sequencing case. The roommate, it turned out, worked at the Broad Institute in Cambridge, Massachusetts, one of the nation's premier biomedical and genomic research centers.

When the Broad scientists had seen the Medical College paper, the roommate explained, they quickly convened a large meeting to discuss pursuing similar work in clinical sequencing.

Partners HealthCare, the largest hospital chain in New England, launched a pilot program to sequence the genomes of select patients and formally introduced sequencing to its hospitals early in 2012. "Across the nation, and spectacularly led by the initial effort in Milwaukee, the use of sequencing is coming into its own," said Robert C. Green, the associate director for research at Partners Personalized Medicine.

Duke University Medical Center introduced a program that employs sequencing to diagnose severely ill infants and imperiled unborn babies with structural defects. The babies are identified as candidates either at their twenty-week ultrasounds or at birth, and parents are given the option of having them sequenced. "I think in the next two to three years this type of activity will be standard operating procedure in hospitals and medical schools across the country," Nicholas Katsanis, the director of Duke's Center for Human Disease Modeling, said in early 2011.

More patient stories appeared, fueling public interest in the new technology. Among the most publicized was that of the Beery twins from California. Misdiagnosed with cerebral palsy, then subjected to years of uncertainty, Alexis and Noah Beery had their exomes sequenced late in 2010, about a year after Nic. Doctors at Baylor College of Medicine pinpointed a previously undiscovered mutation and added a new medication that helped the Beery teenagers, especially Alexis, who'd been suffering from severe breathing difficulties.

"I literally believe it saved Alexis's life," said Retta Beery, their mother.

Other scientists invited Jacob, Worthey, and Dimmock to present talks and seminars describing the Volker case and the system they had established for using genomic medicine in the hospital.

Among the thorniest issues that emerged was the question of whether to provide families with secondary findings—information that's significant to patients but unrelated to the condition that prompted their sequencing. The broad approach of sequencing all of a patient's genes to find the cause of an illness means that doctors might find other mutations linked to as-yet-undiagnosed illnesses. Should doctors tell patients about everything they find, even if the mutations cause a disease for which there is no treatment? It's not simply a question of whether the patient might want to know such things. The information could have consequences for the patient's relatives as well.

Jacob and his team became strong advocates for giving families all of the information they wish to receive. During counseling sessions prior to sequencing they ask patients to decide whether they would want results for "actionable" diseases such as breast cancer and high cholesterol, and also for more dire, incurable conditions such as Parkinson's and Huntington's diseases.

"Your kid could be the next Nic—but we could also find a whole lot of nothing, or a whole lot of other stuff you wish you hadn't known," explained Regan Veith, the Children's Hospital of Wisconsin genetics counselor.

Veith and her colleagues were very clear that patients, not doctors, must make those decisions. "It is our intention if they request that information, we will return it to them," Dimmock told a National Institutes of Health committee in the fall of 2010.

The issue of how much to disclose to families will continue to be debated for a long time, said Kevin Davies, a geneticist, author, and founding editor of *Nature Genetics*. But he agreed with the Wisconsin policy. There is nothing more personal than one's own genome, and it would be "fundamentally wrong" for any doctor to think that he or she has more right to someone's genome than the individual does, Davies said.

"That's just one more area," he said, "where Howard has done this really amazing job of pointing the field in the right direction and saying 'come along and follow me.'" Jacob's lab has set the precedent.

As scientists ranging from Jacob's team in Wisconsin to the NIH Undiagnosed Diseases Program began using genome and exome sequencing more frequently, they became increasingly aware of how extraordinary the Nic Volker story was. Scientists now had a powerful new tool with which to solve the mysteries of rare diseases, but they were only just beginning to learn its strengths and limitations. Sequencing would not yield an answer every time.

In 2013, scientists at Baylor College of Medicine published the first academic paper about a large-scale effort to use sequencing in the clinic. A year later, the Baylor researchers said they had sequenced about 1,700 people with rare diseases and found genetic mutations in about 25 percent of them.

Other sequencing clinics and programs—by 2014 there were roughly twenty in North America—achieved comparable results. Five years after the analysis of Nic Volker's genes, the NIH Undiagnosed Diseases Program had sequenced about four hundred patients with rare illnesses; about seventy had led to a diagnosis, and another sixty or so had yielded a promising genetic suspect.

The caveat was that determining the cause of a person's disease was not the same thing as being able to treat it. Often the answer was that doctors knew what the problem was but were utterly defeated by it. In cases when sequencing led to a diagnosis, only about 30 percent of the time did it affect the clinical care of a patient—and a change in treatment might only improve, rather than cure, a condition. Over time, however, some "failures" may be reversed as scientists learn more about genes and develop new strategies to edit or bypass the harmful actions they set in motion.

By spring 2014, the Medical College and Children's Hospital sequencing clinic had mined the genetic scripts of 436 people, mostly using the exome, or protein-coding segments, as they had for Nic. Like others, the Medical College scientists found that only a quarter of the cases resulted in a diagnosis for a patient whose illness had previously baffled doctors. Yet increasingly, those analyzing genetic scripts found that defeats were sometimes only temporary. As more was learned about the genome from both clinical and research work, the rate of diagnosis rose significantly. Month after month, papers appeared implicating new genes and new mutations in different ailments; all of a sudden, puzzling genetic scripts made sense. The Wisconsin team found that the 25 percent diagnosis rate climbed to almost 40 percent by the eighteen-month mark after a patient had been sequenced.

And in those cases where there was a diagnosis, it was definitive. "We're not talking about a situation where we've found a really interesting variant. This is a slam dunk. This is the answer," said David Bick, the medical director of genetics for Children's Hospital and a professor at the Medical College.

Sequencing was moving with "startling rapidity" into the clinic, and there was a sense among scientists that they were putting together pieces of the jigsaw puzzle of human disease, said Elizabeth Phimister, a deputy editor at *The New England Journal of Medicine*.

At the National Human Genome Research Institute, David Adams, the deputy director of clinical genomics, polled experts around the country in late 2015 and estimated that in the previous twelve months, the exomes of 10,000 to 20,000 patients and their relatives had been sequenced. Stephen Kingsmore, president/CEO of the Rady Pediatric Genomics and Systems Medicine Institute in San Diego, cut the time needed to sequence and analyze infants in the intensive care unit to twenty-six hours; six years earlier it had taken Nic's doctors roughly three months. New technology held the promise of advances yet to come. A British company called Oxford Nanopore Technologies began marketing a sequencing machine in the form of a USB stick that can be plugged into a computer. Although the device was not yet able to sequence an entire human genome, Kevin Davies said he could imagine the day when a blood sample would be taken in a doctor's office and screened within hours for a virus or a condition like Lyme disease.

Successes like Nic Volker and the Beery twins remained rare, but they served to spark the scientific and popular imagination.

Fyodor Urnov, the scientist at Sangamo BioSciences who worked with Jacob's lab to test gene-editing technology in rats, compared the birth of clinical sequencing to the introduction of penicillin, known as the wonder drug of World War II. It was Howard Florey, an Australian pharmacologist, who carried out the first clinical trials of the antibiotic in 1941, after it had sat on the shelf for nearly a decade following its discovery. "Nic is the poster child, and looking at the history of the development of biomedicine, such poster children are incredibly important because they create a 'yes, we can' mind-set in the field," Urnov said.

When Jacob decided to sequence Nic, he challenged medicine's established practice, which favored diagnostic companies such as Myriad Genetics, the longtime patent holder on the BRCA1 and BRCA2 genes. Such companies made handsome profits on single-gene tests.

"It's easier to bill for one test. It's easier to analyze the data for one test," Davies said. The Wisconsin scientists showed that systematically searching every one of Nic's genes could be more cost-effective than relying on a battery of single-gene tests, which in his view amounted to little more than educated guesswork. And they launched a movement to export the new technology beyond the research lab to hospitals around the globe.

As with any landmark achievement, though, there were detractors.

In July 2011, when Worthey addressed a conference in Newport, Rhode Island, critics accused the Medical College scientists of being "cowboys" who had taken the new technology to the clinic before it was ready. They called the sequencing of Nic "a publicity stunt."

That accusation particularly annoyed Worthey, Jacob, and the others, since they had not sought out the publicity they had received in the popular press. In fact, they had shared the details of the sequencing only after being approached in early 2010 by *Milwaukee Journal Sentinel* reporters who were pursuing a tip.

And though the popular press showed interest, the scientific press proved surprisingly more challenging. Worthey and the other Medical College researchers finished drafting the paper on Nic's sequencing in March 2010. It would not be published until eight months later.

"We sent it to five different journals. It took us a year to get it published," Worthey later recalled. One of the reviewers concluded that the work was "not novel enough." Worthey was baffled.

There is a well-established hierarchy of medical and scientific journals, with a handful including *Nature, Science, Cell,* and *The New England Journal of Medicine,* at the top. Then there are what scientists privately call the second-tier and third-tier journals. As they suffered rejections, the Medical College scientists were forced to scale down their expectations for the Nic Volker paper.

One reason they encountered such difficulty is likely that some of the top journals regarded the Yale paper as the initial breakthrough in clinical sequencing. Another reason concerns the categories into which journal articles are slotted. In order to be considered a scientific study, the Medical College paper needed to describe multiple cases, not just one. Because Nic was the first person in the world to be diagnosed with his illness, the paper could not compare his case with others. There were no others.

Medical journals also publish case reports, descriptions of how frontline doctors have treated an unusual, individual patient. The Medical College paper met the definition of a case report, but it then went well beyond it by suggesting—correctly, as it turned out—that exome and genome sequencing could become standard tools in medical practice. In other words, the story the Wisconsin scientists were telling was a unique case study, but it was also the leading edge of a broader trend.

With the movement that Jacob's team launched toward genomic medicine well under way, it is worth asking, for the sake of history, whether skeptics had a point in criticizing the sequencing of Nic Volker. In comments—sometimes private, sometimes in the open—detractors raised two points.

The doctors, they insisted, could have proceeded with Nic's bone marrow transplant without sequencing. Alan Mayer, however, knew how perplexing Nic's ailment was and remembered how firm David Margolis, the transplant expert, had been in insisting on a diagnosis. To go ahead with a transplant without understanding the problem they were trying to fix would have been reckless. Without the information from Nic's exome, there would have been no transplant.

Critics raised a second concern, suggesting that the Medical College scientists had never proved that the mutation they found was the one causing the child's disease. James Verbsky, the immunologist who led the functional studies of Nic's XIAP gene, was aware of this criticism. He remained convinced, however, that his XIAP studies proved that the mutation prevented the gene from doing its vital work, throwing the boy's immune system into a full attack on healthy tissue. And since the transplant, Nic has not developed any further holes in his intestine.

By 2014, much of the initial skepticism had died out and Jacob was being treated as an important figure in genomics. While many of his peers were asked to be on panels at the Personalized Medicine World Conference in Silicon Valley in January, he was invited to give a fifteen-minute solo talk, an indication of his rising reputation. He used the opportunity to discuss what he considered one of the next pressing issues in the field.

Though some still focused on exome sequencing, he argued that patients would be best served if doctors moved straight to sequencing entire genomes. Using Worthey's Carpe Novo software, Jacob's team found that exome sequencing missed, on average, more than two hundred variants known to cause disease; whole-genome sequencing missed just one, Jacob told conference attendees.

Alan Cowley, the Medical College physiology department chairman who staked his career on hiring Jacob, had mixed emotions as he watched his protégé's accomplishments mount. Cowley was proud and enjoyed having had a hand in the scientific achievement. But he was also cautious, worried that the public might expect personalized medicine to take hold more quickly than was possible, particularly for complex diseases. "I think we will get there, but it's not tomorrow," he said.

Complicated cancers, and common diseases such as hypertension, diabetes, and certain mental disorders, are caused by variations in many different genes and are influenced by how the genes and variants interact with one another. Diet, stress, pollution, and other environmental factors also play a role.

It might be decades before DNA sequencing could be effectively and affordably used to predict and treat the most common diseases affecting the most people, Cowley reasoned. Still, since the decoding of the human genome in 2003, Cowley has marveled at the way the exome chip along with leaps in computing speed and capacity have propelled the field beyond expectations.

The rapid progress also has surprised James Watson, whose work a half century ago established the foundation for the Human Genome Project and for Jacob's breakthrough. As he reflected on Jacob's boldness in attacking the crucial challenge of a new era, it reminded him of his own decision many years ago.

"Since I've come out of the South Side of Chicago, which wasn't very affluent, you know, I've always liked the low-odds people who sometimes win," he said. "Often the main thing that makes you successful is just the decision to go ahead and try. The real liberating factor comes from just people willing to try something. DNA sequencing, in some sense, could have been done anywhere.

"In 1951, I saw that X-ray picture of DNA. I was in Naples. I decided I was going for it."

That same spirit drove Alan Mayer when he wrote the letter to Jacob explaining the urgency of Nic's situation. And it drove Jacob when he made the decision to sequence Nic.

In early 2015, Jacob made another decision to go for it, this one involving a new position at the HudsonAlpha Institute for Biotechnology, an organization whose initial funding came from the sale to Invitrogen Corporation of Research Genetics, the company in Alabama that Jacob had turned down almost twenty years earlier. As HudsonAlpha's chief medical genomics officer, Jacob is pursuing an even bigger goal: getting DNA sequencing into every hospital system in the country.

To him, this is the next step in a journey that began with the diagnosis that saved Nic Volker's life. How many Nics? he had asked after the breakthrough. It was a question that assumed the start of a new era.

"I've always believed genomics could change medicine," he says. "Now I know it can."

Acknowledgments

Our former editor at the *Milwaukee Journal Sentinel,* Marty Kaiser, came up with the name "One in a Billion," and from the beginning it felt like the right title for the story of Nic Volker. It certainly sums up our good fortune throughout the reporting and writing of this book.

We will be forever grateful to the Volker family: Sean and Amylynne, their daughters Mariah, Kristen, and Leilani, and most of all, Nic. For five years, this pioneering family allowed us into their lives, patiently answering our endless questions. Amylynne, in particular, was enormously generous in offering her time and access to her CaringBridge online journal, which was the most raw, immediate, and complete chronicle of her son's years in the hospital. Without the Volker family's assistance, this book would not have happened.

We owe a huge debt, too, to the doctors and nurses who cared for Nic and the scientists who sequenced his genetic script. Children's Hospital of Wisconsin and the Medical College of Wisconsin cooperated fully with our reporting, allowing us to watch Nic's story unfold before anyone knew how it would end. Alan Cowley explained how he built a genomics-ready environment. Howard Jacob shared his vision for clinical genomics as well as crucial information about the decision to sequence to Nic Volker. Alan Mayer described Nic's care and the events that led him to approach Jacob about sequencing. David Dimmock, Elizabeth Worthey, and Mike Tschannen explained in detail how they carried out the sequencing and analysis. With admirable candor, James Verbsky described his initial skepticism about sequencing and the functional tests he later carried out to prove it had worked. David Margolis detailed the decision to treat Nic with a cord blood transplant and explained the process. All read portions of the story that became this book. Many others at the hospital and medical center helped in numerous ways, including our media relations contacts Erin Hareng, Maureen Mack, Toranj Marphetia, and Gerry Steele.

A great many scientists shared their expertise, including: George Church, Francis Collins, Richard Gibbs, Eric Green, Mary-Claire King, Eric Lander, Peter Tonellato, Eric Topol, and Nobel Prize winners James Watson and Walter Gilbert. Tom Albert offered invaluable assistance with numerous technical descriptions. Kevin Davies was kind enough to read our manuscript for accuracy and provided a number of wise suggestions.

In another newsroom the story of Nic Volker might have vanished from the front page after a day or two. But editors and former editors Marty Kaiser, George Stanley, Greg Borowski, Thomas Koetting, Becky Lang, Chuck Melvin, and others at the *Milwaukee Journal Sentinel* have spurred the staff to aim high. They gave us the time to report the Nic Volker story and drove us to do it well. Many journalists at the paper read drafts and contributed ideas and encouragement, especially the talented colleagues with whom we shared the 2011 Pulitzer Prize: photographer Gary Porter, graphics editor Lou Saldivar, and videographer Alison Sherwood.

The series might never have made it to the bookstore had Mike Ruby not introduced us to our tenacious agent Flip Brophy, who believed in the project and fought to bring it to the right editor. That editor, Jonathan Cox, was patient, artful, demanding, tireless, and utterly committed to the book. We will always be grateful.

Finally, we owe more than words can say to our families, whose love and sacrifice kept us going. Mark's wife, Mary-Liz, read early drafts and talked through numerous reporting and writing challenges. Their son, Evan, composed and performed the music that inspired Mark through many days of writing.

Kathleen's husband, Bob, and children, Andrew and Emily, were constant in their love and support when their wife and mother was sequestered in the upstairs office during the long stretches of time needed for this project.

Notes

Much of the information for this book came from traditional sources. We interviewed the members of the Volker family and almost all of Nic's doctors. Many of the interviews were recorded, then transcribed. We spoke with Nic's nurses, family friends who came to see him at Children's Hospital of Wisconsin, and friends from church. Also interviewed were many of the key players whose discoveries laid the groundwork for the advent of genomic medicine.

This book made use of written sources, including scores of papers and studies published in scientific journals, and reports from the National Institutes of Health and the National Human Genome Research Institute. We often used curricula vitae, or CVs, as the long, detailed lists of career moves and achievements are called in academia, to verify dates of employment, grant awards, and other information for individual scientists.

Social media were crucial for certain aspects of our reporting. Amylynne Santiago Volker posted frequently on Facebook throughout her son's battle with illness. Even more important was the CaringBridge journal she kept, which was book length and often included details about Nic's treatment, medications, trips to surgery, and total days in the hospital. The journal provided such a thorough and immediate record of Nic's medical odyssey that in some cases it was often more reliable for dates and details than the memories of other sources.

We were fortunate to have a wealth of familiarity with the Medical College of Wisconsin and Children's Hospital of Wisconsin from thirty-seven years of combined coverage of those organizations as reporters at the *Milwaukee Journal Sentinel*.

Chapter 1: At the Threshold

1   *It may turn out:* Watson and Berry, *DNA: The Secret of Life*, 404–5.

2   *James Watson and Francis Crick:* Interview with James D. Watson.

2   *The ability to read our genome:* Interviews with George Church, Francis Collins, David Dimmock, Richard Gibbs, Howard Jacob, Margaret Hamburg, and Francis Collins, "The Path to Personalized Medicine," *New England Journal of Medicine* 363 (2010): 301–4.

4   *This lonely milepost rests:* Interviews with Marjorie Arca, David Dimmock, Bill Grossman, Subra Kugathasan, Alan Mayer, James Verbsky, Amylynne Santiago Volker, Sean Volker; also, Amylynne Santiago Volker's CaringBridge journal.

9   *Taken together, rare diseases:* Office of Rare Diseases, National Institutes of Health.

Chapter 2: Beyond the Four Letters

11   *Their meeting on a beach in La Jolla:* Interviews with Allen Cowley, Howard Jacob.

13   *The march toward decoding:* James Watson and Francis Crick, "Molecular Structure of Nucleic Acids," *Nature* 171 (1953): 737–8; Allan Maxam and Walter Gilbert, "A New Method for Sequencing DNA," *Proceedings of the National Academy of*

Sciences 74 (1977): 560–4; F. Sanger et al., "DNA Sequencing with Chain-terminating Inhibitors," *Proceedings of the National Academy of Sciences* 74 (1977): 5463–7.

13   *When the US government launched:* National Human Genome Research Institute, "The Human Genome Project Completion: Frequently Asked Questions," www.genome.gov/11006943.

14   *Many researchers in physiology and other disciplines:* Interviews with Allen Cowley; Shirley Tilghman, "Lessons Learned, Promises Kept: A Biologist's Eye View of the Genome Project," *Genome Research* 6 (1996): 773.

15   *The Human Genome Project would use:* National Human Genome Research Institute, National Institutes of Health.

16   *He was the son:* Interview with Allen Cowley; "Obituary: Allen Wilson Cowley, M.D.," *Deseret News*, October 22, 2007.

16   *Guyton's lab was building:* Interview with Allen Cowley; John E. Hall et al., "Arthur C. Guyton Obituary," *The Physiologist* 46 (3): 122–124, 2003.

17   *After decades of financial difficulties:* Medical College of Wisconsin history, www.mcw.edu/aboutMCW/HistoryofMCW.htm.

18   *While physiology departments at other schools:* Interviews with Allen Cowley, Richard Katschke, associate vice president, Medical College of Wisconsin Office of Public Affairs.

Chapter 3: Try Me

19   *Jacob had become interested:* Interviews with Dick Jacob, Janet Jacob, Howard Jacob, Lisa Jacob.

21   *Lander knew that interpreting:* Mesirov. *Very Large Scale Computation in the 21st Century,* 138.

22   *In the 1980s, Lander attracted attention:* Eric S. Lander and David Botstein, "Mapping Mendelian Factors Underlying Quantitative Traits Using RFLP Linkage Maps," *Genetics* 121 (1988): 185–99.

22   *Mendel spent about eight years:* Gregor Mendel, "Experiments
       in Plant Hybridization," *MendelWeb*, February 8 and March 8,
       1865, www.mendelweb.org/Mendel.html.

24   *Thirteen days later, the prestigious journal published:* Jacob
       et al. "Genetic Mapping of a Gene Causing Hypertension in
       the Stroke-prone Spontaneously Hypertensive Rat," *Cell* 67
       (1991): 213–24.

25   *Cowley patiently courted Jacob:* Interviews with Allen Cowley,
       Howard Jacob, Lisa Jacob, Richard Roman.

27   *The Medical College would provide:* Interviews with Allen
       Cowley, Howard Jacob; Carolyn B. Alfvin, "Golden Lab," *Mil-
       waukee Magazine*, January 1996.

Chapter 4: The Pied Piper

30   *The work would result in a paper:* Monika Stoll et al., "A
       Genomic-Systems Biology Map for Cardiovascular Function,"
       Science 294 (2001): 1723–1726.

30   *They would publish another paper:* Theodore A. Kotchen et
       al., "Identification of Hypertension-related QTLs in African
       American Sib Pairs," *Hypertension* 40 (2002): 634–9.

30   *A third paper in 2005:* P. Hamet et al., "Quantitative Founder-
       Effect Analysis of French Canadian Families Identifies Specific
       Loci Contributing to Metabolic Phenotypes of Hypertension,"
       American Journal of Human Genetics 76 (2005): 815–32.

30   *their 1997 Banbury Conference:* Cowley, "Genomics to Physi-
       ology and Beyond: How Do We Get There?," *The Physiologist*
       40 (1997): 205.

31   *The center's goal was deceptively simple:* Interviews with How-
       ard Jacob, Jozef Lazar, Peter Tonellato; "Report of the Genetics
       Initiative Committee," September 28, 1998 (Jacob chaired this
       committee).

31   *won a five-year, $7.3 million grant:* Interviews with Peter Ton-
       ellato; Peter Tonellato, curriculum vitae.

31   *By 2001, the number of lab rats:* Tour of Medical College of Wisconsin Biomedical Resource Center; interviews with Jozef Lazar, Joseph Thulin.

33   *the goal of sequencing an entire human genome for $1,000:* Davies, *The $1,000 Genome,* 9.

34   *Jacob presented a bold timeline:* Interviews with Howard Jacob; Jacob's draft plan "Institute for Personalized Medicine," July 10, 2003.

Chapter 5: An Extraordinary Patient

35   *It begins innocently enough:* Interviews with Amylynne Santiago Volker, Mariah Volker, Leilani Varese Volker, Sean Volker.

39   *The wound from the abscess never heals:* Interviews with Amylynne Santiago Volker, Sean Volker.

40   *Crohn's very seldom strikes:* Interviews with James Verbsky, Bill Grossman, Alan Mayer.

42   *She and Sean understand that behind the doctors' questions:* Interviews with Amylynne Santiago Volker, Sean Volker.

43   *treating some sixty thousand patients a year:* Children's Hospital of Wisconsin.

44   *Kugathasan has treated:* Interview with Subra Kugathasan.

44   *He puts Nic on a Crohn's medication:* Interviews with Amylynne Santiago Volker; her CaringBridge journal.

44   *She grew up in a rural area:* Interviews with Marjorie Arca.

Chapter 6: Diagnostic Odyssey

46   *On her first examination of Nic:* Interviews with Marjorie Arca, Amylynne Santiago Volker; her CaringBridge journal.

47   *About 30 percent of them:* Interview with William Gahl at the Office of Rare Diseases, National Institutes of Health.

47   *One is William Grossman:* Interviews with Bill Grossman.

48   *The child is so sick:* Interviews with Bill Grossman.

48   *In Nic's case, the cells are failing:* Interview with James Verbsky.

48   *Is the child's disease causing:* Interview with Bill Grossman.

49   *Even doctors who specialize:* Interviews with Subra Kugath-asan, Alan Mayer.

49   *By late June 2007:* Amylynne Santiago Volker's CaringBridge journal.

50   *Doctors give Nic blood transfusions:* Amylynne Santiago Volker's CaringBridge journal.

50   *In the next few years:* Interviews with Sean Volker, Amylynne Santiago Volker.

Chapter 7: Spiders on the Ceiling

51   *Two days after Nic pronounces himself cured:* Interviews with Sean Volker, Amylynne Santiago Volker.

52   *At Children's Hospital, doctors quickly confirm:* Interviews with George Hoffman, Amylynne Santiago Volker, Sean Volker.

52   *Nic's hallucinations mean that:* Interview with George Hoffman.

53   *Some nights doctors tell her:* Interviews with Amylynne Santiago Volker; her CaringBridge journal.

53   *A temperature of 108 is almost always fatal:* Interview with George Hoffman.

53   *Septic shock patients typically wax and wane:* Interview with George Hoffman.

54   *Some nights she curls up beside Nic in bed:* Interviews with Amylynne Santiago Volker.

54   *"Nicholas has a couple of breathing scares":* Amylynne Santiago Volker's CaringBridge journal; interview with George Hoffman.

54   *"Nicky is in Day 5":* Amylynne Santiago Volker's CaringBridge journal.

54   *Amylynne's father, the doctor, encourages her:* Interviews with Amylynne Santiago Volker.

55   *Nic receives numerous transfusions:* Interview with George Hoffman.

55   *any weight gain:* Amylynne Santiago Volker's CaringBridge journal.

55   *"He is looking good":* Ibid.

Chapter 8: Big Steps and Fast

57   *an ambulance would deliver:* Mark Johnson and Kawanza Newson, "Sole Survivor: A Journey of Faith and Medicine," *Milwaukee Journal Sentinel*, June 18, 2005, www.jsonline.com/news/health/47465427.html.

58   *Bob Kliegman has made good:* Interviews with Howard Jacob, Bob Kliegman.

59   *it becomes the focus:* Howard Jacob, curriculum vitae.

59   *crucial to many of the 150-plus scientific papers:* Howard Jacob, his curriculum vitae.

60   *The first gene-editing tool:* Interviews with Jozef Lazar, Howard Jacob.

60   *His search for a faster:* Interviews with Howard Jacob, Fyodor Urnov.

61   *a paper that will be published:* Aron M. Geurts et al., "Knockout Rats via Embryo Microinjection of Zinc-Finger Nucleases," *Science* 325 (2009): 433.

62   *A $100 million attempt to identify:* National Human Genome Research Institute.

62   *The next key step:* National Human Genome Research Institute.

62   *The study, known as the Wellcome Trust Case Control Consortium:* P. R. Burton et al., "Genome-wide Association Study of 14,000 Cases of Seven Common Diseases and 3,000 Shared Controls," *Nature* 447 (2007): 667–8.

62   *they would be dogged:* Nicholas Wade, of the *New York Times*, summed up the questions around the International HapMap

and the genome-wide association studies in a story published June 12, 2010, titled "A Decade Later, Genetic Map Yields Few New Cures," www.nytimes.com/2010/06/13/health /research/13genome.html?pagewanted=all&_r=0.

63 *it was to a community hospital:* Interview with Bob Kliegman.

63 *Broeckel led Children's first genomic partnership:* Interviews with Howard Jacob, Bob Kliegman.

64 *A new $140 million Children's Research Institute building:* Children's Hospital of Wisconsin Researcher Resources, www.chw .org/medical-professionals/research/researcher-resources/; interview with Joe Lazar.

64 *The hospital's pediatrics department:* Interview with Bob Kliegman.

64 *completed the sequence of his entire genome:* Davies, *The $1,000 Genome,* 16.

64 *Venter declared:* Craig Venter, "The Richard Dimbleby Lecture 2007: Dr. J. Craig Venter—A DNA-Driven World," May 12, 2007, www.bbc.co.uk/pressoffice/pressreleases/stories/2007/12 _december/05/dimbleby.shtml.

64 *Jonathan Rothberg, the founder of 454 Life Sciences:* Davies, *The $1,000 Genome,* 16.

65 *a next-generation sequencing machine:* Interview with Jaime Wendt Andrae.

Chapter 9: Patient X

66 *After work:* Interviews with Bill Grossman, James Verbsky, Michael Stephens.

67 *A more general theory among the doctors:* Interviews with Bill Grossman, James Verbsky.

67 *In this scenario:* Interviews with David Margolis.

68 *In February 2008:* Interviews with Amylynne Santiago Volker; her CaringBridge journal.

68   *"the immunologist believes again"*: Amylynne Santiago Volker's CaringBridge journal.

69   *the Volkers are having problems:* Interviews with Sean Volker, Amylynne Santiago Volker.

69   *He weighs just 19 pounds:* Amylynne Santiago Volker's Caring-Bridge journal.

69   *Doctors tell Amylynne:* Interviews with Amylynne Santiago Volker.

70   *The hospital is known across the country:* Interviews with Ted Denson, Amylynne Santiago Volker.

70   *As best the Cincinnati doctors can tell:* Interview with Ted Denson.

70   *The truth is that his bottom and intestinal system:* Interview with Alan Mayer.

Chapter 10: No More Secrets

73   *One day a nurse:* Interview with Amylynne Santiago Volker.

74   *With tens of thousands of patients admitted each year:* Children's Hospital of Wisconsin 2014 Fact Sheet, www.chw.org/~/media/Files/About/CHW_Facts_0414.pdf.

75   *As a child, Amylynne:* Interviews with Amylynne Santiago Volker.

76   *she would come to the split-level:* Interviews with Amylynne Santiago Volker, including a tour she provided of the town of Monroe to the authors.

77   *Sean is not entirely happy:* Interviews with Sean Volker, Amylynne Santiago Volker.

78   *Leilani, who is visiting her brother:* Interviews with Leilani Varese Volker, Amylynne Santiago Volker.

79   *In 2008 alone:* Amylynne Santiago Volker's CaringBridge journal; interviews with Marjorie Arca.

Chapter 11: Survivors

81   *But the other doctors Mayer consults:* Interviews with Alan Mayer.

81   *Mayer's sympathy for Nic is balanced:* Ibid.

82   *His father survived for a year and a half:* Ibid., *Entombed,* 1–206.

82   *Because he showed a gift with numbers:* Interviews with Alan Mayer.

84   *The gastrointestinal system is complex:* Ibid.

84   *At Mass General, he worked:* Ibid.

85   *Vanderbilt and Washington Universities:* Ibid.

86   *The odds of receiving funding:* Ibid.; information from the National Institutes of Health.

87   *The turning point in his deliberations:* Ibid.; Amylynne Santiago Volker's CaringBridge journal.

88   *There are only two ways:* Interviews with Alan Mayer.

89   *Mayer begins by speaking:* Interviews with Michael Stephens, Alan Mayer.

89   *He knows Amylynne loves her son:* Interviews with Alan Mayer.

90   *"As far as they can tell":* Amylynne Santiago Volker's CaringBridge journal.

Chapter 12: The Dragon

91   *But in Nic's case:* Interviews with Marjorie Arca.

92   *The wound cleanings are full-blown procedures:* Interviews with Amylynne Santiago Volker, Marjorie Arca.

92   *Wound-cleaning days follow:* Interviews with Marjorie Arca, Amylynne Santiago Volker.

93   *Nic and Amylynne are in the midst:* Interviews with Amylynne Santiago Volker; her CaringBridge journal.

93   *The boy knows where the operating room is:* Interviews with Amylynne Santiago Volker.

93   *At night, he sleeps cradling a bag of Bagel Bites:* Ibid.

93   *there are other days:* Interviews with Amylynne Santiago Volker; her CaringBridge journal.

94   *They think of Nic:* Interviews with Mariah and Leilani Varese Volker.

94   *Mariah, the oldest:* Interview with Leilani Varese Volker.

95   *When the family's six-year-old Nissan Altima:* Interviews with Amylynne Santiago Volker.

95   *One night he comes home:* Interview with Leilani Varese Volker.

95   *At job sites, sometimes:* Interviews with Sean Volker.

95   *In three months, Nic has spent:* Interviews with Amylynne Santiago Volker; her CaringBridge journal.

96   *Amylynne did not want this:* Amylynne Santiago Volker's CaringBridge journal.

96   *They try the honey:* Interviews with Marjorie Arca, Amylynne Santiago Volker.

97   *Finally, on April 22:* Ibid.

97   *Nic endures a collapsed lung, fevers:* Interviews with Marjorie Arca, Amylynne Santiago Volker; excerpts from Amylynne Santiago Volker's CaringBridge journal.

98   *He receives transfusions of granulocytes:* Interviews with Alan Mayer.

98   *At times, Mayer does not even feel:* Ibid.

99   *The dragon follows him home each day:* Ibid.

Chapter 13: The Genome Joke

100   *Mayer has tried various medications:* Interviews with Alan Mayer.

100   *Still, the doctor observes:* Ibid.

101   *If he loses his ileum:* Ibid.

101   *But when he tests his theory:* Interviews with James Verbsky.

101   *When a bone marrow transplant is not:* Interview with Alan Mayer.

102 *Amylynne tries to lift their spirits:* Interviews with Amylynne Santiago Volker.

102 *In desperation, he goes to see:* Interviews with Alan Mayer, David Margolis.

103 *At this point, late June 2009:* Interviews with Eric Green; PubMed searches.

103 *"Dear Howard," he writes:* Email from Alan Mayer to Howard Jacob, provided by Mayer.

Chapter 14: Why Are We Here?

106 *He shares his parents' Roman Catholic faith:* Interviews with Howard, Dick, and Janet Jacob.

107 *Lazar's boss in Baltimore:* Interviews with Jozef Lazar.

107 *"This is our chance":* Ibid.

108 *Elizabeth Worthey has already read it:* Interviews with Elizabeth Worthey.

108 *It is their first set of human genome data:* Ibid.

109 *The prevailing view:* Interviews with David Dimmock, James Verbsky, Howard Jacob.

110 *A week after sending the email:* Interviews with Alan Mayer.

111 *The cost of the human work alone:* Interview with Eric Green.

111 *The first human genome, now known:* Interview with Lisa Brooks.

112 *The private sequencing effort:* Venter, *A Life Decoded*, 286.

113 *Within a day:* Interviews with Howard Jacob.

114 *In the conference room:* Interviews with Howard Jacob, David Dimmock, Elizabeth Worthey, Alan Mayer.

115 *But Verbsky believes:* Interview with James Verbsky.

116 *In his experience, the very best genetic tests:* Interview with David Dimmock.

117 *In 2008, Jacob told Tschannen:* Interviews with Mike Tschannen.

118 *One day, not long after the meeting:* Interviews with Howard Jacob, Mike Tschannen.

Chapter 15: Unknown Territory

120 *The advantage of such wide approaches:* Interviews with Howard Jacob, David Dimmock.

122 *Sometimes the script tells a story:* Ibid.

122 *Two of the doctors:* Interview with David Margolis; Amylynne Santiago Volker's CaringBridge journal.

122 *Such is Nic's condition:* Interviews with Alan Mayer, Amylynne Santiago Volker; Volker's CaringBridge journal.

123 *When the discussion of a possible bone marrow transplant:* Interviews with Amylynne Santiago Volker.

Chapter 16: I'm Going to Tell You a Story

125 *this will be Howard Jacob's last board meeting:* Interviews with Howard Jacob, Brian Curry, Jeff Harris.

127 *Osterberger thinks of her own three children:* Interview with Jolene Osterberger.

127 *Nic has exhausted his $2 million:* Interviews with Amylynne Santiago Volker.

128 *rare diseases cut a vast swath of misery:* Office of Rare Diseases, National Institutes of Health.

128 *Roche, which has been building:* Roche announced in March 2007 that it would buy 454 Life Sciences and in June 2007 that it would buy NimbleGen Systems.

129 *the FDA had met in July:* Letters on June 10, 2010, and November 22, 2013, from the U.S. Food and Drug Administration to 23andMe: www.fda.gov/downloads/MedicalDevices/Resources forYou/Industry/UCM215240.pdf and www.fda.gov/ICECI/En forcementActions/WarningLetters/2013/ucm376296.htm.

129 *Jacob has succeeded in raising about half:* Interviews with Howard Jacob.

Chapter 17: Fine White Threads

130 *The sample contains material:* Interviews with Mike Tschannen, Gwen Shadley.

131 *There is a machine:* Interviews with Mike Tschannen.

131 *Shadley regards her work as a special calling:* Interview with Gwen Shadley.

132 *There was a set of routine steps:* Ibid.

133 *Tschannen does not tell her:* Interviews with Mike Tschannen, Gwen Shadley.

133 *The work with Nic's blood:* Interviews with Gwen Shadley.

134 *Tschannen picks up one of the vials:* Interview with Mike Tschannen.

134 *Technicians begin by breaking:* Ibid.

135 *The chip they use:* Interviews with Tom Albert.

136 *The Roche technicians add tiny beads:* Interviews with Mike Tschannen and Tom Albert.

137 *In Wisconsin, Elizabeth Worthey:* Interviews with Elizabeth Worthey, David Dimmock.

137 *As they construct the master list:* Interviews with Howard Jacob, Elizabeth Worthey, David Dimmock.

138 *Facing thousands of possible mutations:* Interviews with Elizabeth Worthey.

Chapter 18: Thousands of Suspects

140 *They destroy her son's immune system:* Interviews with Alan Mayer, David Margolis.

140 *The child's fever hovers:* Excerpt from Amylynne Santiago Volker's CaringBridge journal.

140 *One day, he looks:* Interviews with Amylynne Santiago Volker.

141 *Working with Medical College software developers:* Interviews with Elizabeth Worthey.

141 *Some changes in the script:* Interviews with David Dimmock.

141 *One database called dbSNP:* Interviews with David Dimmock, Elizabeth Worthey.

142 *What Worthey needs:* Interviews with Elizabeth Worthey.

142 *When the results arrive:* Ibid.

142 *16,124 differences between Nic's script:* Interviews with David Dimmock, Elizabeth Worthey.

143 *For more than a year:* Interviews with David Dimmock, Elizabeth Worthey.

143 *More than a decade earlier:* Interviews with David Dimmock.

144 *At the end of September:* Interviews with Mike Tschannen.

145 *At this point, each segment:* Ibid.

145 *Worthey and Dimmock describe the thirty two:* Interviews with David Dimmock, Elizabeth Worthey.

146 *She thinks of this as her "hot list":* Interviews with Elizabeth Worthey.

147 *The first consequence:* Ibid.

147 *On November 16, 2009:* Email from Elizabeth Worthey.

147 *But since then, a new paper has appeared:* Interviews with Alan Mayer, James Verbsky, Elizabeth Worthey; Andreas Krieg et al., "XIAP Mediates NOD Signaling via Interaction with RIP2," *Proceedings of the National Academy of Sciences* 106 (2009): 14524–9.

148 *She and Dimmock study the effect:* Interviews with David Dimmock, Elizabeth Worthey.

149 *It's XIAP:* Interviews with Alan Mayer, Elizabeth Worthey.

Chapter 19: The Culprit

150 *In one specific position:* Interviews with David Dimmock, Alan Mayer, Elizabeth Worthey.

151 *That amino acid is the 203rd:* Ibid.

151 *Without enough XIAP protein:* Interviews with James Verbsky.

152 *By 2009, there are only a dozen:* Interview with Elizabeth Worthey.

152 *So Worthey widens the genetic net:* Interviews with David Dimmock, Howard Jacob, Elizabeth Worthey.

152 *Dimmock contacts still other scientists:* Ibid.

152 *Researchers around the world:* National Human Genome Research Institute.

153 *to their surprise, the sequencing has revealed:* Interviews with David Dimmock, Alan Mayer.

154 *It sometimes seems:* Interviews with Amylynne Santiago Volker; her CaringBridge journal.

155 *On a Friday afternoon:* Interviews with Amylynne Santiago Volker, Alan Mayer.

156 *For that reason:* Interviews with Amylynne Santiago Volker; her CaringBridge journal.

157 *Somewhere around 3 to 4 percent:* Interview with Marlene Zuk, biological sciences professor, University of Minnesota.

157 *But by early December:* Interviews with Amylynne Santiago Volker; her CaringBridge journal.

158 *If the doctor is right:* Interviews with Alan Mayer.

158 *a few days before Christmas:* Amylynne Santiago Volker's CaringBridge journal.

160 *each of these single-gene tests:* Interviews with James Verbsky, Bill Grossman.

160 *In order to prove:* Interviews with James Verbsky.

Chapter 20: Faith and Doubt

161 *Three years of medical emergencies:* Interviews with Amylynne Santiago Volker, Sean Volker, Leilani Varese Volker; Amylynne Santiago Volker's CaringBridge journal.

162 *The test in Seattle:* Interviews with David Dimmock.

162 *Dimmock and genetics counselor Regan Veith:* Interviews with David Dimmock, Regan Veith.

162 *Veith knows there is an unavoidable tendency:* Interview with Regan Veith.

163 *"You carry the mutation," he says:* Interviews with David Dimmock, Amylynne Santiago Volker.

164 *In January, Dimmock, Mayer, and:* Interviews with David Dimmock, Alan Mayer.

165 *But the process carries:* Interviews with David Margolis.

165 *Amylynne begins considering:* Interviews with Amylynne Santiago Volker.

165 *While they determine where to go:* Amylynne Santiago Volker's CaringBridge journal.

166 *One flash point is a trip:* Interviews with Alan Mayer, Amylynne Santiago Volker.

166 *In the three and a half years:* Amylynne Santiago Volker's CaringBridge journal.

167 *The endoscopy reveals:* Interviews with Alan Mayer, Amylynne Santiago Volker.

168 *In June, a few days before:* Interviews with Amylynne Santiago Volker; her CaringBridge journal.

Chapter 21: The Smell of Creamed Corn

170 *Now, though, for the first time:* Interviews with Amylynne Santiago Volker.

171 *Over the next hour:* Ibid.

171 *Margolis is a great believer:* Interview with David Margolis.

172 *"Mommy," he says:* Interviews with Amylynne Santiago Volker; her CaringBridge journal.

172 *Margolis has decided:* Interview with David Margolis.

173 *Nic explains to his mother:* Interviews with Amylynne Santiago Volker.

173 *At 2:17, with a storm brewing outside:* Video of the transplant provided by the Volker family.

174 *The days that follow:* Interviews with Amylynne Santiago Volker; her CaringBridge journal.

174 *There are lengthy discussions:* Interviews with David Dimmock, Alan Mayer.

175 *Eight days after the transplant:* Interviews with David Margolis.

175 *Sores form on his mouth, throat, tongue, gut:* Amylynne Santiago Volker's CaringBridge journal.

176 *By day 20, the adenovirus is advancing:* Amylynne Santiago Volker's CaringBridge journal.

177 *The medical team hustles:* Interviews with David Margolis.

177 *Day 47, August 30:* Amylynne Santiago Volker's CaringBridge journal.

178 *The big day arrives at last:* Reporting by the authors for the *Milwaukee Journal Sentinel.*

Chapter 22: Inheritance

180 *He thunders around the Monona house:* Interviews with Amylynne Santiago Volker; her CaringBridge journal.

182 *He fights off another bout of sepsis:* Ibid.

182 *He sometimes wishes:* Interviews with Amylynne Santiago Volker.

182 *he has the DNA of two distinct people:* Interviews with David Margolis.

182 *The most alarming legacy of Nic's transplant:* Ibid.

183 *After Nic completes eight weeks:* Interviews with Amylynne Santiago Volker.

184 *The couple grew apart:* Ibid.

184 *In the years immediately following Nic's crisis:* Interviews with Amylynne Santiago Volker, Leilani Varese Volker.

185 *During her sophomore year of high school:* Interviews with Leilani Varese Volker.

185 *she talks of becoming:* Interviews with Leilani Varese Volker.

185 *But no other family members:* Interviews with Amylynne Santiago Volker, Leilani Varese Volker.

186 *Sean is aware that his script:* Interviews with Amylynne Santiago Volker.

186 *The odds are 50 percent:* Interviews with David Dimmock.

187 *Nic remembers a surprising amount:* Interviews with Nic Volker and Amylynne Santiago Volker.

187 *For the first time in his young life:* Interviews with Amylynne Santiago Volker.

188 *For family celebrations:* Interviews with Amylynne Santiago Volker and Leilani Varese Volker.

Chapter 23: Come Along and Follow Me

189 *Months after Nic's discharge:* Interviews with Howard Jacob.

190 *When she learned:* Interview with Mary Claire King.

191 *Doctors had used exome:* Murim Choi et al., "Genetic Diagnosis by Whole Exome Capture and Massively Parallel DNA Sequencing," *Proceedings of the National Academy of Sciences* 106 (2009): 19096–101, and interviews with Tom Albert.

191 *Turkish doctors had contacted:* Interview with Richard Lifton.

193 *Nic, Collins suggested:* Francis Collins, "Faces in the Genome," *Science* 331 (2011): 546.

193 *At a conference in Florida:* Interview with Mike Tschannen.

194 *Partners HealthCare:* Interview with Robert C. Green.

194 *Duke University Medical Center introduced:* Interview with Nicholas Katsanis.

194 *Alexis and Noah Beery:* Interviews with James Lupski, Retta Beery.

195 *Jacob and his team became strong advocates:* Interviews with David Dimmock, Howard Jacob; David Dimmock, comments

at the 23rd meeting of the Secretary's Advisory Committee on Genetics, Health, and Society, October 5, 2010.

195 *Veith and her colleagues:* Interviews with Regan Veith.

195 *The issue of how much:* Interviews with Kevin Davies.

196 *A year later, the Baylor researchers:* Interview with Richard Gibbs.

196 *Five years after the analysis of Nic Volker's genes:* Interview with William Gahl.

196 *By spring 2014:* Interviews with David Bick, Howard Jacob.

198 *In July 2011:* Interviews with Elizabeth Worthey.

201 *Alan Cowley, the Medical College physiology department chairman:* Interviews with Alan Cowley.

201 *also has surprised James Watson:* Interview with James Watson.

202 *In early 2015, Jacob made another decision:* Interviews with Howard Jacob.

Sources

Interviews (conducted 2010–2015)

Thomas Albert, Marjorie Arca, Alan Attie, Michael Bamshad, Tara Bell, David Bick, Leslie Biesecker, Mark Boguski, Lisa Brooks, David Brow, Daniel Burgess, George Church, Jolene Clark, Robert Clarke, Francis Collins, Alan Cowley, Heather Cunliffe, Brian Curry, Kevin Davies, Brennan Decker, Ted Denson, David Dimmock, Dian Donnai, Fred Emani, Greg Feero, William Gahl, Arupa Ganguly, Richard Gibbs, David Gifford, Walter Gilbert, David Goldstein, Eric Green, Robert C. Green, Bill Grossman, Hakon Hakonarson, Tina Hambuch, Jeff Harris, Daniel Helbling, George Hoffman, Joshua Hyman, Dick Jacob, Howard Jacob, Janet Jacob, Lisa Jacob, Nicholas Katsanis, Richard Katschke, Mary-Claire King, Robert Kliegman, Subra Kugathasan, Eric Lander, Jozef Lazar, Charles Lee, Richard P. Lifton, Dennis Lo, James Lupski, David Margolis, Alan Mayer, Amy McGuire, Michael F. Murray, Vinodh Narayanan, Jolene Osterberger, Thomas Pearson, Elizabeth Phimister, Frank Pitzer, Reed Pyeritz, Jonathan Ravdin, Cynthia Rayford, John Raymond, Kevin Regner, Heidi Rehm, Gregory Rice, Richard Roman, Stan Rose, Derrick Rossi, Jack Routes, Jeffery Schloss, Rebecca Selzer, John Seman, Gwen Shadley, Jay Shendure, Jan Smith, Michael Stephens, Mark Stevenson, Keith Stewart, Michael Sussman, Joseph Thulin,

Peter Tonellato, Eric Topol, Mike Tschannen, Steve Turner, Kevin Ulmer, Julie Upton, Fyodor Urnov, Regan Veith, James Verbsky, Amylynne Santiago Volker, Leilani Varese Volker, Mariah Volker, Nic Santiago Volker, Sean Volker, Daniel Von Hoff, James Watson, Jaime Wendt Andrae, Eric Wieben, Marilou Wijdicks, Liz Worthey, Kim Zebell, Marlene Zuk.

Books

Davies, Kevin. *The $1,000 Genome: The Revolution in DNA Sequencing and the New Era of Personalized Medicine.* New York: Free Press, 2010.

Mayer, Bernard. *Entombed.* Miami, FL: Aleric Press, 1994.

McElheny, Victor K. *Drawing The Map of Life: Inside the Human Genome Project,* New York: Basic Books, 2010.

Mesirov, Jill P. *Very Large Scale Computation in the 21st Century.* Philadelphia: Society for Industrial and Applied Mathematics, 1991.

Mukherjee, Siddhartha. *The Emperor of All Maladies: A Biography of Cancer.* New York: Simon and Schuster, 2010.

Venter, J. Craig. *A Life Decoded. My Genome: My Life.* New York: Penguin Group, 2007.

Watson, James. *The Double Helix: A Personal Account of the Discovery of the Structure of DNA.* New York: Atheneum, 1968.

Watson, James, and A. Berry. *DNA: The Secret of Life.* New York: Random House, 2003.

Papers

2013

Lazar, Jozef, et al. "SORCSI Contributes to the Development of Renal Disease in Rats and Humans." *Physiological Genomics* 45: 720–8.

Rangel-Filho, Artur, et al. "Rab38 Modulates Proteinuria in Model of Hypertension-Associated Renal Disease." *Journal of the American Society of Nephrology* 24: 283–92.

2012

Hedge, Madhuri R. "Marching Towards Personalized Genomic Medicine." *The Journal of Pediatrics* 162: 10–11.

Need, Anna C., et al. "Clinical Application of Exome Sequencing in Undiagnosed Genetic Conditions." *Journal of Medical Genetics.* doi:10.1136/jmedgenet-2012-100819.

Phimister, Elizabeth, et al. "Realizing Genomic Medicine." *New England Journal of Medicine* 366: 757–9.

Zerbino, D., et al. "Integrating Genomes." *Science* 336: 179–82.

2011

Bainbridge, Matthew, et al. "Whole-Genome Sequencing for Optimized Patient Management." *Science Translational Medicine* 3, issue 87, 87re3.

Bick, David, and David Dimmock. "Whole Exome and Whole Genome Sequencing." *Current Opinion in Pediatrics* 23: 594–600.

Collins, Francis. "Faces in the Genome." *Science* 331: 546.

Evans, James, et al. "Deflating the Genomic Bubble." *Science* 331: 861–2.

Feero, W. Gregory, et al. "Genomics and the Eye." *New England Journal of Medicine* 364: 1932–42.

Green, Eric, and Mark Guyer. "Charting a Course for Genomic Medicine from Base Pairs to Bedside." *Nature* 470: 204–13.

Tonellato, Peter, et al. "A National Agenda for the Future of Pathology in Personalized Medicine: Report of the Proceedings of a Meeting at the Banbury Conference Center on Genome-era Pathology, Precision Diagnostics, and Preemptive Care: A Stakeholder Summit." *American Journal of Clinical Pathology* 135: 663–5.

Worthey, E. A., et al. "Making a Definitive Diagnosis: Successful Clinical Application of Whole Exome Sequencing in a Child with Intractable Inflammatory Bowel Disease." *Genetics in Medicine* 13: 255–62.

Zierhut, Heather. "How Inclusion of Genetic Counselors on the Research Team Can Benefit Translational Science." *Science Translational Medicine* 3(74), 74cm7.

2010

Annes, Justin, et al. "Risks of Presymptomatic Direct-to-Consumer Genetic Testing." *New England Journal of Medicine* 363: 1100–1.

Arensburger, Peter, et al. "Sequencing of *Culex quinquefasciatus* Establishes a Platform for Mosquito Comparative Genomics." *Science* 330: 86–8.

Ashley, Euan, et al. "Clinical Assessment Incorporating a Personal Genome." *The Lancet* 375: 1525–35.

Cho, Mildred, and David Relman. "Synthetic 'Life,' Ethics, National Security, and Public Discourse." *Science* 329: 38–9.

Dietz, Harry. "New Therapeutic Approaches to Mendelian Disorders." *New England Journal of Medicine* 363: 852–63.

Dowell, Robin, et al. "Genotype to Phenotype: A Complex Problem." *Science* 328: 469.

Feero, W. Gregory. "Genomic Medicine—an Updated Primer." *New England Journal of Medicine* 362: 2001–11.

Hamburg, Margaret, and Francis Collins. "The Path to Personalized Medicine." *New England Journal of Medicine* 363: 301–4.

Kumar, Rajesh, et al. "Genetic Ancestry in Lung-Function Predictions." *New England Journal of Medicine* 363: 321–30.

Lo, Y. M. Dennis, et al. "Maternal Plasma DNA Sequencing Reveals the Genome-wide Genetic and Mutational Profile of the Fetus." *Science Translational Medicine* 2: 61.

Lupski, James, et al. "Whole-Genome Sequencing in a Patient with Charcot-Marie-Tooth Neuropathy." *New England Journal of Medicine* (2010): 1181–91.

McClellan, Jon, and Mary-Claire King. "Genetic Heterogeneity in Human Disease." *Cell* 141: 210–17.

Ng, Sara, et al. "Exome Sequencing Identifies MLL2 Mutations as a Cause of Kabuki Syndrome." *Nature Genetics* 42: 790–3.

Ng, Sarah B., et al. "Exome Sequencing Identifies the Cause of a Mendelian Disorder." *Nature Genetics* 42: 30–5.

Ormond, Kelly, et al. "Challenges in the Clinical Application of Whole-Genome Sequencing." *The Lancet* 375: 1525–35.

Roach, Jared, et al. "Analysis of Genetic Inheritance in a Family Quartet by Whole-Genome Sequencing." *Science* 328: 636–9.

Rotini, Charles, and Lynn Jorde. "Ancestry in the Age of Genomic Medicine." *New England Journal of Medicine* 363 (2010): 1551–8.

Samani, Nilesh. "The Personal Genome—the Future of Personalized Medicine?" *The Lancet* 375: 1497–8.

Scherer, Stephen, and Autism Genome Project Consortium. "Functional Impact of Global Rare Copy Number Variation in Autism Spectrum Disorders." *Nature* 466: 368–72.

Sobreira, Nara, et al. "Whole-Genome Sequencing of a Single Proband Together with Linkage Analysis Identifies a Mendelian Disease Gene." *PLoS Genetics*. DOI: 10.1371/journal.pgen .1000991.

Sudmant, Peter, et al. "Diversity of Human Copy Number Variation and Multicopy Genes." *Science* 330: 641–6.

Wang, K., et al. "Interpretation of Association Signals and Identification of Causal Variants from Genome-wide Association Studies." *American Journal of Human Genetics* 86: 1–13.

Zaghloul, Norann, et al. "Functional Analyses of Variants Reveal a Significant Role for Dominant Negative and Common Alleles in Oligogenic Bardet-Biedl Syndrome." *Proceedings of the National Academy of Sciences* 107: 10602–7.

2009

Choi, Murim, et al. "Genetic Diagnosis by Whole Exome Capture and Massively Parallel DNA Sequencing." *Proceedings of the National Academy of Sciences* 106: 19096–101.

Geurts, Aron M., et al. "Knockout Rats via Embryo Microinjection of Zinc-Finger Nucleases." *Science* 325: 433.

Glocker, Erik-Oliver, et al. "Inflammatory Bowel Disease and Mutations Affecting the Interleukin-10 Receptor." *New England Journal of Medicine* 361: 2033–45.

Krieg, Andreas, et al. "XIAP Mediates NOD Signaling via Interaction with RIP2." *Proceedings of the National Academy of Sciences* 106: 14524–9.

Manolio, Teri. "Finding the Missing Heritability of Complex Diseases." *Nature* 461: 747–53.

Nagy, Noemi, et al. "The Proapoptotic Function of SAP Provides a Clue to the Clinical Picture of X-linked Lymphoproliferative Disease." *Proceedings of the National Academy of Sciences* 106: 11966–71.

Ng, Sara, et al. "Targeted Capture and Massively Parallel Sequencing of 12 Human Genomes." *Nature* 461: 272–6.

Watson, James. "Living with My Personal Genome." *Personalized Medicine* 6: 607.

2008

Lee, Charles, and Cynthia Morton. "Structural Genomic Variation and Personalized Medicine." *New England Journal of Medicine* 358: 740–1.

Schloss, Jeffery. "How to Get Genomes at One Ten-thousandth the Cost." *Nature Biotechnology* 26: 1113–15.

Shendure, Jay, and Hanlee Ji. "Next-Generation DNA Sequencing." *Nature Biotechnology* 26: 1135–45.

2007

Albert, Thomas, et al. "Direct Selection of Human Genomic Loci by Microarray Hybridization." *Nature Methods* 4: 903–5.

Burton, P. R., et al. "Genome-wide Association Study of 14,000 Cases of Seven Common Diseases and 3,000 Shared Controls." *Nature* 447: 667–78.

Hodges, Emily, et al. "Genome-Wide in situ Exon Capture for Selective Resequencing." *Nature Genetics* 39: 1522–7.

Lucast, Erica. "Informed Consent and the Misattributed Paternity Problem in Genetic Counseling." *Bioethics* 21: 41–50.

Mischler, Matthew, et al. "Epstein-Barr Virus-Induced Hemophagocytic Lymphohistiocytosis and X-linked Lymphoproliferative Disease: A Mimicker of Sepsis in the Pediatric Intensive Care Unit." *Pediatrics* 119: 1212–18.

Okou, David, et al. "Microarray-based Genomic Selection for High-Throughput Resequencing." *Nature Methods* 4: 907–9.

2006

Mischler, Matthew, et al. "Epstein-Barr Virus–Induced Hemophagocytic Lymphoproliferative and X-Linked Lymphoproliferative Disease: A Mimicker of Sepsis in the Pediatric Intensive Care Unit." *Pediatrics* 119: 1212–18.

Sjöblom, Tobias, et al. "The Consensus Coding Sequences of Human Breast and Colorectal Cancers." *Science* 314: 268–74.

2005

Bonham, V. L., et al. "Race and Ethnicity in the Genome Era." *American Psychologist* 60: 9–15.

Check, Erika. "Patchwork People." *Nature* 437: 1084–6.

Hamet, P., et al. "Quantitative Founder-Effect Analysis of French Canadian Families Identifies Specific Loci Contributing to Metabolic Phenotypes of Hypertension." *American Journal of Human Genetics* 76: 815–32.

Lazar, Jozef, et al. "Impact of Genomics on Research in the Rat." *Genome Research* 15: 1717–28.

2004

Gibbs, Richard A., et al. "Genome Sequence of the Brown Norway Rat Yields Insights into Mammalian Evolution." *Nature* 428: 493–521.

2003

Collins, Francis, et al. "A Vision for the Future of Genomic Research." *Nature* 422: 835–47.

Moreno, Carol, et al. "Genomic Map of Cardiovascular Phenotypes of Hypertension in Female Dahl S Rats." *Physiological Genomics* 15: 243–57.

2002

Broeckel, Ulrich, et al. "A Comprehensive Linkage Analysis for Myocardial Infarction and Its Related Risk Factors." *Nature Genetics* 30: 210–14.

Freedman, Barry I., et al. "Linkage Heterogeneity of End-Stage Renal Disease on Human Chromosome 10." *Kidney International* 62: 770–4.

Guttmacher, Alan, and Francis Collins. "Genomic Medicine—a Primer." *New England Journal of Medicine* 347: 1512–20.

Hunt, Steven C., et al. "Linkage of Creatinine Clearance to Chromosome 10 in Utah Pedigrees Replicates a Locus for End-Stage Renal Disease in Humans and Renal Failure in the Fawn-hooded Rat." *Kidney International* 62: 1143–8.

Kotchen, Theodore A., et al. "Identification of Hypertension-related QTLs in African American Sib Pairs." *Hypertension* 40: 634–9.

Suzuki, Miwako, et al. "Genetic Modifier Loci Affecting Survival and Cardiac Function in Murine Dilated Cardiomyopathy." *Circulation* 105: 1824–29.

2001

Lander, Eric S., et al. "Initial Sequencing and Analysis of the Human Genome." *Nature* 409: 860–921.

Stoll, Monika, et al. "A Genomic-Systems Biology Map for Cardiovascular Function." *Science* 294: 1723–6.

2000

Kissebah, Ahmed H., et al. "Quantitative Trait Loci on Chromosomes 3 and 17 Influence Phenotypes of the Metabolic Syndrome." *Proceedings of the National Academy of Sciences* 97: 14476–83.

Shiozawa, Masahide, et al. "Evidence of Gene-Gene Interactions in the Genetic Susceptibility to Renal Impairment After Unilateral

Nephrectomy." *Journal of the American Society of Nephrology* 11: 2068–78.

1998

Knapik, Ela W., et al. "A Microsatellite Genetic Linkage Map for Zebrafish (*Danio rerio*)." *Nature Genetics* 18: 338–43.

1997

Cowley, Allen W., Jr. "Genomics to Physiology and Beyond: How Do We Get There?" *The Physiologist* 40: 205–11.

1996

Brown, Donna M., et al. "Renal Disease Susceptibility and Hypertension Are Under Independent Genetic Control in the Fawn-hooded Rat." *Nature Genetics* 12: 44–51.

Tilghman, Shirley M. "Lessons Learned, Promises Kept: A Biologist's Eye View of the Genome Project." *Genome Research* 6: 773–80.

1995

Jacob, Howard J., et al. "A Genetic Linkage Map of the Laboratory Rat, *Rattus norvegicus*." *Nature Genetics* 9: 63–9.

1994

Lander, Eric S., and Nicholas J. Schork. "Genetic Dissection of Complex Traits." *Science* 265: 2037–47.

1991

Jacob, Howard J., et al. "Genetic Mapping of a Gene Causing Hypertension in the Stroke-prone Spontaneously Hypertensive Rat." *Cell* 67: 213–24.

1989

Rapp, John P., et al. "A Genetic Polymorphism in the Renin Gene of Dahl Rats Cosegregates with Blood Pressure." *Science* 243: 542–4.

1988

Lander, Eric S., and David Botstein. "Mapping Mendelian Factors Underlying Quantitative Traits Using RFLP Linkage Maps." *Genetics* 121: 185–99.

Patterson, Andrew H., et al. "Resolution of Quantitative Traits into Mendelian Factors by Using a Complete Linkage Map of Restriction Fragment Length Polymorphisms." *Nature* 335: 721–6.

1977

Maxam, Allan, and Walter Gilbert. "A New Method for Sequencing DNA." *Proceedings of the National Academy of Sciences* 74: 560–4.

Sanger, F., et al. "DNA Sequencing with Chain-terminating Inhibitors." *Proceedings of the National Academy of Sciences* 74: 5463–7.

1953

Watson, James, and Francis Crick. "Molecular Structure of Nucleic Acids." *Nature* 171: 737–8.

1865

Mendel, Gregor. "Experiments in Plant Hybridization." *MendelWeb*, February 8 and March 8, 1865, www.mendelweb.org/Mendel.html.

Index

About the Authors

Mark Johnson is a health and science reporter at the *Milwaukee Journal Sentinel*, where he has worked since 2000. He was a member of the *Journal Sentinel* team that won the Pulitzer Prize for explanatory reporting on the Nic Volker story in 2011. He is also a three-time finalist for the Pulitzer Prize and has won numerous other awards for his reporting. He lives with his wife and son in Fox Point, Wisconsin. He was also guitarist for the punk band, The Bloody Stumps.

Kathleen Gallagher is a business reporter at the *Milwaukee Journal Sentinel*, where she has worked since 1993. She was a member of the *Journal Sentinel* team that won the Pulitzer Prize for explanatory reporting on the Nic Volker story in 2011. She was part of a team that won the 2006 Inland Press Association award and has won many other awards for her reporting. She lives with her husband and two children in Wauwatosa, Wisconsin.